# Hydrogels in Tissue Engineering

# Hydrogels in Tissue Engineering

Special Issue Editor

**Esmaiel Jabbari**

MDPI • Basel • Beijing • Wuhan • Barcelona • Belgrade

MDPI

*Special Issue Editor*
Esmaiel Jabbari
University of South Carolina
USA

*Editorial Office*
MDPI
St. Alban-Anlage 66
Basel, Switzerland

This is a reprint of articles from the Special Issue published online in the open access journal *Gels* (ISSN 2310-2861) from 2016 to 2018 (available at: http://www.mdpi.com/journal/gels/special_issues/poly_tissue_engineering)

For citation purposes, cite each article independently as indicated on the article page online and as indicated below:

LastName, A.A.; LastName, B.B.; LastName, C.C. Article Title. *Journal Name* **Year**, *Article Number*, Page Range.

**ISBN 978-3-03897-121-4 (Pbk)**
**ISBN 978-3-03897-122-1 (PDF)**

Cover image courtesy of Esmaiel Jabbari.

# Contents

# About the Special Issue Editor

**Esmaiel Jabbari**, PhD, Professor of Chemical and Biomedical Engineering. Prof. Jabbari completed his Ph.D. in Chemical Engineering at Purdue University under the mentorship of Professor Nicholas A. Peppas. He is Tenured Full Professor of Chemical and Biomedical Engineering at the University of South Carolina. His research interest is the development of multi-cellular tissue models for skeletal tissue engineering with spatiotemporal morphogen delivery. He began his independent career as an Assistant Professor of Biomedical Engineering at Mayo Clinic upon completion of his training at Monsanto and Rice University. Prof. Jabbari received the Berton Rahn Award in 2012 from AO Foundation and the Stephen Milam Award in 2008 from the Oral and Maxillofacial Surgery Foundation. He was elected Fellow of AIMBE in 2013. He is the author of >250 research articles and has given >100 invited lectures. He serves as the Academic Editor for PLOS ONE and Associate Editor for Gels.

# Preface to "Hydrogels in Tissue Engineering"

Hydrogels are hydrophilic solid materials that do not dissolve in aqueous or physiological medium. Hydrogels hold a large quantity of water in their structure, to the extent that the diffusivity of molecules in hydrogels is close to liquids not solids. The diffusivity of oxygen, glucose, and a typical protein like albumin in a polyethylene glycol (PEG) hydrogel is, on average and depending on the solid content, molecular weight, and extent of gelation, 100 $\mu$m2/s, 50 $\mu$m2/s, and 10 $\mu$m2/s, respectively, which is only an order magnitude lower than its respective diffusivity in water. Conversely the diffusivity of oxygen in hydrophobic soft materials like natural rubber is 1000 times lower than in water with glucose and proteins having much lower diffusivity. As a result of their solid form and high permeability to oxygen, nutrients, and proteins, hydrogels are used extensively in medicine to replace soft, as well as hard, tissues.

Otto Wichterle and Drahoslav Lim, in a seminal paper published in Nature (O. Wichterle, D. Lím, Hydrophilic gels for biological use, Nature 185, 1960, 117–118), argued that hard hydrophobic plastics, due to a mismatch in mechanical properties and the slow leaching of toxic low molecular weights compounds, pose a serious biocompatibility problem when in contact with natural biological tissues. Wichterle and Lim proposed that a biocompatible material should (1) have a molecular structure affording the desired water content, (2) be inert to normal biological processes, and (3) have permeability to metabolites. Wichterle and Lim went on to invent (patented in 1959) the first hydrogel based on poly(hydroxyl ethyl methacrylate) or p(HEMA) and demonstrated its usefulness as a soft contact lens. However, it did not become commercially available until 1971 when Bausch and Lomb launched the first FDA-approved soft contact lens. During the 1960s and 1970s, the use of hydrogels in drug delivery was extensively explored by pharmaceutical scientists and engineers, which led to the development of environmentally-sensitive hydrogels. Today, pH-sensitive hydrogels like poly(acrylic acid) are used commercially in many oral dosage forms to prevent drug release in the acidic environment of the stomach and allow release in the higher pH of duodenum and colon. Today, hydrogels are being used in many trades and products including, but not limited to, drug delivery, diapers, water storage micro-reservoir in agriculture, cosmetics, plastic surgery, vaccination, cancer therapy, water purification, cultivation of micro-organisms, and tissue repair and regeneration.

The last few decades have witnessed the rapid rise in research and development for the use of hydrogels in soft tissue replacement, repair, and regeneration. In this regard, hydrogels based on natural biopolymers like collagen, hyaluronic acid, alginate, and chitosan, due to their excellent biocompatibility and non-toxic degradation products, are extensively used in tissue repair. The use of synthetic hydrogels in tissue replacement is constrained by our lack of understanding of the fate and toxicity of degradation by-products of synthetic gels after implantation and our limited understanding of the fate and function of cells in contact with engineered hydrogels. Therefore, there is a pressing need to develop novel hydrogels with controllable degradation with non-toxic degradation products that support the function and maturation of the implanted cells to specified lineage and phenotype. Articles in this volume focus on the rationale for the design of hydrogels for tissue regeneration. This includes printing biphasic hydrogels for the regeneration of load-bearing skeletal tissues, polyampholyte hydrogels to prevent the microbial fouling of tissue constructs, bioresponsive hydrogels for cell delivery, hydrogels that mimic the tissue extracellular matrix for cell encapsulation, hydrogels for high-throughput screening of the factors related to the cell microenvironment, and peptide-conjugated hydrogels for cell adhesion and signal transduction.

Finally, I would like to extend my deepest appreciation to all contributing authors whose expert contributions made the publication of this Special Issue possible. I would also like to express my deepest appreciation to the editorial team, especially Ms. Jiao Li at MDPI for encouragement, technical guidance, editing, and publication of this Special Issue.

**Esmaiel Jabbari**
*Special Issue Editor*

*gels*

Editorial

# Hydrogels for Cell Delivery

Esmaiel Jabbari

Biomimetic Materials and Tissue Engineering Laboratory, University of South Carolina,
Columbia, SC 29208, USA; jabbari@cec.sc.edu; Tel.: +01-803-777-8022

Received: 15 June 2018; Accepted: 29 June 2018; Published: 2 July 2018

Hydrogels have a three-dimensional crosslinked molecular structure which absorb large quantities of water and swell in a physiological environment. Hydrogels are a class of polymers made from hydrophilic repeat units that interact with water molecules by hydrogen bonding, polar and ionic interaction to take up water many times the initial polymer weight. Further, the polymer chains in the hydrogel are linked via crosslinks to form an infinite network to prevent dissolution of the polymer chains in an aqueous medium. Hydrogels can be natural or synthetic. Due to their high water content, oxygen molecules, nutrients, peptides, proteins, ribonucleic acid (RNA) and deoxyribonucleic acid (DNA) biomolecules diffuse readily through hydrogels. Further, cells immobilized in hydrogels maintain their viability and function. As a result of these benefits, hydrogels are used extensively in medical applications for replacement, repair, and regeneration of soft biological tissues. There are >8000 references to hydrogels in PubMed and >15,000 in Web of Science search engines.

Recently, hydrogels have been used as a matrix for delivery of cells and morphogens to the site of injury in regenerative medicine. Natural as well as synthetic hydrogels are used in tissue replacement, repair, and regeneration. Natural hydrogels can be derived from plants or animals. Plant-derived hydrogels include polysaccharide-based agarose, alginate, and carboxymethyl cellulose. Animal-derived hydrogels include polysaccharide-based, such as hyaluronic acid, and protein-based, such as collagen, gelatin, chitosan, and fibrin. In particular, injectable and in-situ hardening hydrogels functionalized with photocrosslinkable moieties are very attractive for repairing or regenerating irregularly-shaped tissue injuries using minimally-invasive arthroscopic procedures. In that approach, a suspension of therapeutic cells, morphogens, and growth factors in a functionalized hydrogel precursor solution is injected through a catheter to the injury site guided by imaging. After injection, the precursor solution is hardened or gelled by shinning ultraviolet or visible light enabled catheter.

More recently, hydrogels are being used as bioinks for printing cells, morphogens, and growth factors such that the spatial organization of the printed cells and growth factors mimic that of the target tissue. The hydrogel ink in these cellular constructs serves as an extracellular glue to maintain dimensional ability and provide mechanical strength to the construct. The hydrogel also provides ligands for specific interactions between the cell surface receptors and the extracellular matrix (ECM) guide cellular events like adhesion, migration, mitosis, differentiation, maturation, and protein expression. Multiple printing heads are used to print tissue constructs with many cell types and growth factors.

The articles in this Special Issue provide exemplary reviews and research works related to the use of hydrogels in tissue engineering and regenerative medicine. Although cells encapsulated in hydrogels maintain their viability and function, the high water content significantly reduces the hydrogel's mechanical strength. As a result, hydrogels unaided cannot be used a matrix for regeneration of load-bearing tissues such as bone. To mitigate this issue, Kumar and collaborators describe in their article titled "A Bioactive Hydrogel and 3D Printed Polycaprolactone System for Bone Tissue Engineering" the development of a novel hard–soft biphasic construct with a gyroid geometry by 3D printing. In this approach, a stiff poly($\varepsilon$-caprolactone) (PCL) polymer was used to print the hard phase of the construct in a gyroid geometry, whereas a combination of alginate and gelatin was used to print the soft phase as a carrier for osteoblast progenitor cells. The gyroid geometry of the hard phase increased the volume of the soft phase which, in turn, increased cell loading and the extent of osteogenesis.

A major complication of cellular tissue constructs is microbial fouling after implantation. Individually there are viable options for sterilization of biomaterials, growth factors, and cells. However, complete sterilization of cells, growth factors, and biomaterials collectively in a tissue construct is complicated, even with the use of anti-bacterial and anti-fungal agents. Therefore, strategies that can reduce microbial fouling can significantly enhance their suitability in clinical applications. In that regard, Yu and collaborators describe in their review titled "Polyampholyte Hydrogels in Biomedical Applications" the properties of polyampholyte hydrogels and their non-fouling characteristics. Polyampholytes are an interesting class of hydrogels that possess both positive and negatively charged units in their structure. The interaction between the positive and negatively charged units imparts anti-fouling properties to the hydrogel which can be exploited in tissue engineering applications.

Natural hydrogels are widely used as a carrier for cells in tissue engineering because they contain sequences of amino acids that interact with cell surface receptors to guide cell function and expression. However, it is difficult to tailor the multitude of ligand–receptor interactions in natural matrices to a particular application in regenerative medicine. Further, natural hydrogels suffer from batch-to-batch variability in composition, limited thermal and mechanical stability, and relatively fast and uncontrolled enzymatic degradation. Conversely, synthetic hydrogels have tunable physical and mechanical properties for a wide range of applications in medicine, but they lack instructive interactions with the encapsulated cells. Therefore, there is a need to develop novel synthetic approaches to modify hydrogels with cell-adhesive ligands. Cipolla, Russo and collaborators in "Bioresponsive Hydrogels: Chemical Strategies and Perspectives in Tissue Engineering" and Varghese and collaborators in "Hydrogels as Extracellular Matrix Analogs" describe strategies and approaches to produce functional, cell-responsive hydrogels for applications in regenerative medicine.

In regenerative medicine, a mixture of growth factors as well as many ligand receptor interactions, physical and mechanical factors are involved in differentiation and maturation of progenitor cells to a specific lineage. Therefore, there is a need to develop high-throughput techniques to screen for these factors within a 3D tissue culture system. The review by Dr. Smith Callahan titled "Combinatorial Method/High Throughput Strategies for Hydrogel Optimization in Tissue Engineering Applications" highlights the strengths and disadvantages of design of experiment, arrays and continuous gradients and fabrication challenges for hydrogel optimization in tissue engineering applications.

The interaction of receptors on the cell surface with ECM ligands starts a cascade of signaling from the cell membrane to the cell cytoplasm and the nucleus to activate/deactivate genes of interest. The gene activation in turn leads to protein expression and secretion of the desired ECM and tissue regeneration. Although cell–ECM interactions have been extensively studied in 2D culture system, more work is needed to understand signal transduction in biomimetic 3D cultures with cells encapsulated in hydrogels. Ventre and Netti in "Controlling Cell Functions and Fate with Surfaces and Hydrogels: The Role of Material Features in Cell Adhesion and Signal Transduction" review signal transduction for cells in hydrogels that captures features of the natural cellular environment, such as dimensionality, remodeling and matrix turnover.

*gels*

MDPI

Review

# Controlling Cell Functions and Fate with Surfaces and Hydrogels: The Role of Material Features in Cell Adhesion and Signal Transduction

Maurizio Ventre [1,2] and Paolo A. Netti [1,2,*]

[1] Department of Chemical, Materials and Industrial Production Engineering and Interdisciplinary Research Centre on Biomaterials, University of Naples Federico II, P.le Tecchio 80, 80125 Napoli, Italy; maventre@unina.it

[2] Center for Advanced Biomaterials for Health Care@CRIB, Istituto Italiano di Tecnologia, L.go Barsanti e Matteucci 53, 80125 Napoli, Italy

* Correspondence: nettipa@unina.it; Tel.: +39-081-768-2408

Academic Editor: Esmaiel Jabbari
Received: 22 January 2016; Accepted: 1 March 2016; Published: 14 March 2016

**Abstract:** In their natural environment, cells are constantly exposed to a cohort of biochemical and biophysical signals that govern their functions and fate. Therefore, materials for biomedical applications, either *in vivo* or *in vitro*, should provide a replica of the complex patterns of biological signals. Thus, the development of a novel class of biomaterials requires, on the one side, the understanding of the dynamic interactions occurring at the interface of cells and materials; on the other, it requires the development of technologies able to integrate multiple signals precisely organized in time and space. A large body of studies aimed at investigating the mechanisms underpinning cell-material interactions is mostly based on 2D systems. While these have been instrumental in shaping our understanding of the recognition of and reaction to material stimuli, they lack the ability to capture central features of the natural cellular environment, such as dimensionality, remodelling and degradability. In this work, we review the fundamental traits of material signal sensing and cell response. We then present relevant technologies and materials that enable fabricating systems able to control various aspects of cell behavior, and we highlight potential differences that arise from 2D and 3D settings.

**Keywords:** cell adhesion; surface patterning; hydrogel; mechanotransduction

## 1. Introduction

For a long time, cell-culturing substrates, like glass, plastic and metal, were considered as passive supports. In these systems, soluble biochemical supplements were regarded as key players in affecting cell fate and functions. However, a growing body of experimental evidence has come to light in the recent past and has clearly demonstrated that the chemical-physical properties of the scaffolding materials can be as effective as the soluble biochemical signals [1]. This should not be surprising, since each and every cell is constantly exposed to a multitude of signals *in vivo* that can be biochemical and biophysical in nature. In fact, cells can recognize and respond to mechanical forces of the surrounding environment, gradients of ligands and the topography of the tissues in which they reside [2]. Analogously, material substrates will invariably display signals to cells either deliberately or in an 'unintentional' manner. In other words, materials intrinsically possess their own stiffness, the distribution of adhesion signals; even what we consider a flat surface might display a topography at the nanoscale. Signals displayed by materials can influence a broad spectrum of cellular behaviors, such as adhesion spreading, migration, proliferation and differentiation [3,4]. Despite the sheer number of examples, only a few molecular mechanisms involved in the transduction of material stimuli in biological responses have recently been clarified [5–7]. This notwithstanding, a thorough understanding of the complex, molecular interplays occurring between material signals and cell response would bring in novel design concepts to engineer instructive materials able to control cell fate and functions in a deterministic manner. The practical benefits arising from such knowledge could be tremendous, since it can lead to the development of effective tissue-engineered products, tissue models to study development and pathologies *in vitro* and platforms for drug testing and discovery.

A large body of literature concerning the effects of material stimuli on cell behavior was focused on two-dimensional (2D) substrates that were instrumental in shaping our knowledge on the biochemical transduction of material signals. However, the effective translation of these findings in a clinical context requires the development of three-dimensional (3D) structures that better reproduce a physiological environment. In particular, tissue engineering and regenerative medicine failed in having a dramatic impact on modern clinics, despite their undeniable potentialities. This is mainly caused by a lack of knowledge on the effects of exogenous stimuli and in particular those presented by culturing materials, in the generation of fully-functional tissues *in vitro* or *in vivo*. This becomes particularly relevant in the case of stem cells that are very sensitive to micro-environmental signals [8]. In fact, signals presented to stem cells in their niche ultimately dictate fate and functions, *i.e.*, whether they have to remain quiescent, proliferate or differentiate [9]. In this context, one of the greatest challenges is to develop materials able to display a set of stimuli that tightly control stem cell behavior. This requires designing and fabricating perfectly-controlled physical/chemical environments in which the effect of specific material signals on cell functions can be precisely assessed. Developments in material science and related technologies, such as micro- and nano-fabrication and polymer functionalization, can be particularly useful to achieve this task. The modulation of a broad range of material features can be achieved in 2D setups with consolidated processes. However, tailoring the biochemical/biophysical characteristics of 3D environments requires much more sophisticated approaches. It has to be pointed out that the complexity in controlling cell behavior in 3D does not simply depend on the 'added' dimensionality. As will be soon clear, in 3D, cells perceive material signals differently from what happens in 2D. Furthermore, 3D substrates intrinsically possess additional features, not usually observed in 2D, like degradability or the possibility of undergoing extensive structural remodelling, which ultimately affect cell behavior. Hydrogels proved to be particularly useful in the context of cell behavior control through material features [10]. In fact, they possess chemical/physical characteristics that make them versatile platforms in which stiffness, porosity, bioactivity and degradability can be variously modulated.

In this work, we present and discuss some of the recent and most relevant findings concerning how material features can affect cell adhesion. We then analyze why modulating the adhesion event is important and how to achieve this with material patterning techniques. Finally, we provide examples on controlling cell functions and fate with specifically-engineered systems.

## 2. Mechanics of Cell Adhesion Formation on 2D or 3D Material Systems

Intuitively, the perception of material signals by cells requires some sort of contact followed by a probing phase. In fact, specialized molecular machineries are activated whenever the environmental conditions are permissive for a cell to adhere and spread on a substrate. More specifically, focal adhesion (FA) and the actomyosin cytoskeleton are the structures that play a fundamental role in adhering to and probing the extracellular environment [11,12]. They also provide the mechanical connection with which cells can exert forces to the ECM and *vice versa*: ECM transmits stress and strain to the cell cytoplasm. Several different types of macromolecules constitute FAs. Among these, integrins, transmembrane receptors, specifically engage ligands on the extracellular space, whereas proteins from the cytoplasmic side may exert a signaling (like focal adhesion kinase (FAK) and paxillin) or mechanical functions (like talin, vinculin, actinin and zyxin) [13]. Interestingly, the activity and dynamics of many adhesion molecules appear to be force dependent, for which contractile forces generated by the actin fibers can induce conformational changes that ultimately trigger signaling pathways [14,15]. The presence of certain ligands, the ways these are displayed by the extracellular space, along with their mobility are all factors that affect FA formation and maturation. Integrin clustering is an essential feature for the maturation of stable FAs [16]. Too few or sparse ligands might impair this process and halt the downstream signaling pathways [17]. Additionally, weakly-bound ligands or ligands tethered to flexible structures can be remodeled by the contractile forces exerted by the cell, and this can also affect cell response [18].

The concepts discussed so far are valid both *in vivo* and *in vitro*. In the latter case, materials need to be functionalized in order to display adhesive signals to cells. A broad range of chemical strategies and manipulation technologies have been developed and optimized so far in order to control cell adhesion events. Several works dating back to the early 1990s focused on modulating adhesion events affecting cell functions, such as spreading, migration and proliferation [19–21]. Diverse chemical functionalization strategies and fabrication technologies have been developed so far to precisely control the biochemical/biophysical features of the culturing substrate in order to direct cell behavior.

Generally, the modulation of the cell adhesion events, and the cell response thereof, has been widely investigated in 2D setups. In this context, inorganic materials (glass, metallic alloys) or synthetic polymers (predominantly hard polystyrene (PS), polycaprolactone (PCL) or soft polydimethylsiloxane (PDMS), polyacrylamide (PAM)) have been largely used. Synthetic polymers proved to be particularly useful owing to their intrinsic versatility in allowing biochemical functionalization or the fine modulation of their mechanical properties in a broad range of stiffness. For instance, by simply changing the polymer/crosslink ratio, PAM hydrogels and PDMS elastomers can cover up to three orders of magnitude of Young's modulus spanning from a few kPa up to MPa [22]. Additionally, many hard and rigid polymers are compatible with various micro- and nano-fabrication technologies, which allow embossing complex structures on their surfaces. Historically, adsorption of adhesive proteins (fibronectin, collagen, gelatin, vitronectin, laminin) has been routinely performed to make glass or synthetic materials bioactive. This however results in a poor control on ligand positioning and stability, and this becomes particularly relevant when a weakly-bound ligand layer experiences extensive cell-mediated traction forces. In this case, extensive ligand remodelling might occur, making it difficult to relate the cell response to the initial bioactive properties of the material surface (Figure 1) [23,24].

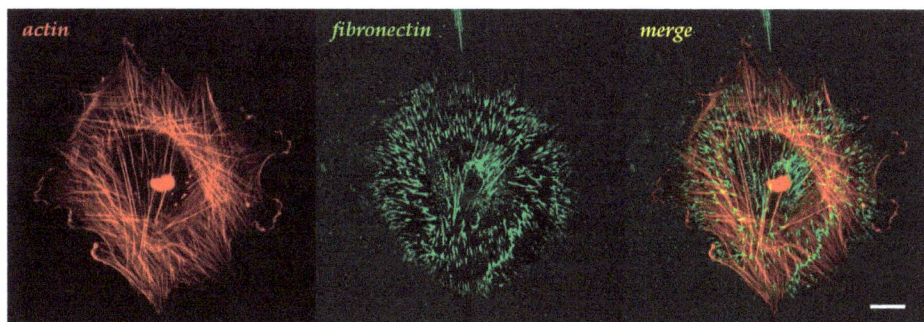

**Figure 1.** Confocal micrograph showing the effect of cell-generated forces on physisorbed fibronectin. MC3T3 preosteoblasts cultivated for 12 h on a nanograted, O2 plasma-treated PDMS substrate. Fibronectin (10 μm/mL) undergoes extensive remodelling caused by contractile forces. Fibronectin compaction is observed at both ends of actin fibers. Note how fibronectin smears follow the actin direction and leave a dark halo upon compaction. Actin is stained with Tetramethylrhodamine B isothiocyanate-phalloidin (red); fibronectin is stained by immunofluorescence (green). Scale bar: 20 μm.

Furthermore, hydrophobic materials can denature physisorbed proteins, and this can affect the actual concentration of ligands presented to cells [25]. Covalent conjugation of proteins on synthetic materials allows gaining a better control over ligand stability and presentation. An enormous variety of chemical routes has been reported in the literature concerning the binding of biomolecules on surfaces, either with or without spacers. Most popular strategies involve glutaraldehyde, carbodiimide [26], sulfosuccinimidyl 6-(4'-azido-2'-nitrophenylamino)hexanoate (*i.e.*, sulfo-SANPAH) cross-linking [27] and the biotin-avidin binding system [28,29]. Yet, handling natural biomolecules to functionalize substrates can be expensive and time consuming. Furthermore, proteins can undergo denaturation or degradation as a result of the chemical treatments necessary for the coupling [30]. More recently, the use of peptide sequences that specifically interact with integrins has become a popular method to control cell adhesion on material surfaces or within scaffolds, owing to their increased stability towards chemical treatments. Examples of short peptide ligands include DGEA, RGD (derived from collagen), IKVAV, RGD, YIGSR (laminin), REDV and RGDS (fibronectin) [31]. RGD is certainly one of the most used and studied sequences, and several studies tracing back to the early 1990s investigated the density of RGD necessary to promote cell spreading and adhesion. Massia and Hubbell found that a density of 1 fmol/cm$^2$ of RGD is sufficient for cell spreading on glass surfaces, whereas 10 fmol/cm$^2$ are sufficient for focal contacts and stress fiber formation [19]. These figures strongly depend on the type of material substrate, since higher amounts of RGD peptides are generally required to achieve cell adhesion [32]. This seems to be related to the nature of the flexible linkers that connect the ligand to the surface; linkers might not provide the correct signal display or an effective mechanical feedback to cells upon contraction [33]. Furthermore, the chemical/physical properties of the surface may alter the effectiveness of ligand display. Additionally, the extracellular domain of integrins projects out of the membrane by ~10 nm, and it is likely that this is the maximum distance that allows for integrin-ligand engagement [34]. If the cell membrane cannot accommodate recesses on the material surface, then nanometric roughness on the material surface, or strata deeper than 10 nm in functionalized hydrogels, can in principle make ligands not readily accessible to the integrins.

While 2D setups possess undeniable advantages, like simple functionalization strategies, direct accessibility to the material regions to be functionalized, no resistance to nutrient transport and suitability to live examination with high magnification lenses, they cannot recapitulate the more physiologically-relevant, but complex 3D architectures found *in vivo*. The control of the biochemical/biophysical features of 3D environments requires the development and implementation of more complex processes. The 3D porous scaffolds used in tissue engineering applications are usually characterized by a pore size in the 100–500-mm range [35]. Within this range, cell behavior is affected by pore curvature [36]; additionally cells gradually fill up the pores and therefore do not perceive the same physical environment as the one sensed initially [37]. Nanofibrous electrospun mats might provide a microenvironment that is morphologically similar to native ECM; however, the modulation of the mechanical properties usually results in a modification of fibril diameter, pore size and bioactivity [38,39]. Conversely, polymeric and biopolymeric hydrogels not only provide cells with an *in vivo*-like 3D environment, but allow a fine tuning of the biochemical, microstructural and mechanical features through consolidated chemical/physical routes.

Early examples of the use of hydrogels in cell biology concern fibrillar natural gels as collagen and fibrin. These are constituted by polypeptides that self-assemble in the form of microfibrils that intertwine in a 3D network (Figure 2a).

The gelation process occurs in mild conditions, thus allowing direct encapsulation of cells. Fibrils naturally display ligands for cell adhesion; therefore, additional functionalizations are generally not required. Despite these positive characteristics, natural fibrillar hydrogels are very compliant, and their structural mechanical and bioactive properties cannot be modulated independently. For instance, increasing the protein concentration results in an increase of gel stiffness, but adhesivity and porosity are also affected. This makes it difficult to assess the role of a specific hydrogel feature on the observed cellular response. However, the increase of modulus obtainable in this manner is marginal. Chemical crosslinking with glutaraldehyde has also been frequently applied, but unreacted molecules are extremely toxic. Other methods involving non-enzymatic glycation or genipin were used to effectively crosslink gels with minimal cytotoxic effects [40,41]. Matrigel is another natural hydrogel widely used for cell cultures. Its components, primarily laminin and collagen IV, are extracted from Engelbreth-Holm-Swarm mouse tumors [42]. Cells cultivated on or within Matrigel are able to recapitulate some morphogenetic processes that eventually lead to self-organized systems displaying striking similarity to native tissues and organs [43]. Also for Matrigel, the mechanical and structural properties cannot be tuned straightaway, and owing to its natural origin, batch-to-batch variability could occur.

(a)

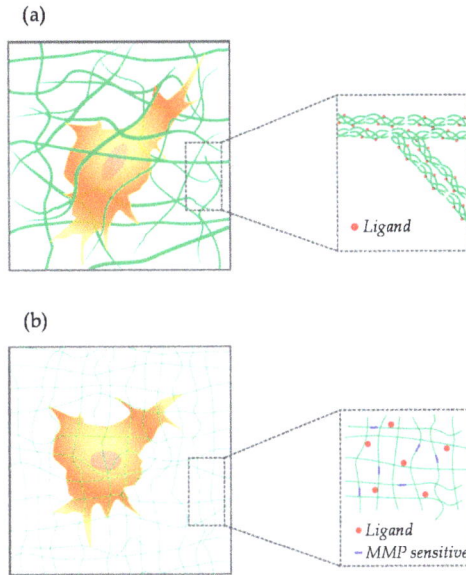

(b)

**Figure 2.** Schematic of cells encapsulated in hydrogels. (**a**) Natural fibrillar hydrogel: proteins self-assemble in the form of fibrils that form an entangled mass surrounding the cells. Fibrils constitutively display ligand motifs for cell attachment. (**b**) Polymeric hydrogel (synthetic or saccharidic): ligands need to be conjugated to the polymeric backbone, as well as degradable domains, to allow cells to adhere and spread.

Hydrogels whose constituents possess a "simple" chemical structure that can be precisely modified are steadily gaining popularity as 3D ECM analogues. These gels can be either of natural (agarose, alginate, hyaluronan) or synthetic origin (polyethylene glycol (PEG), poly(vinyl alcohol) (PVA), PAM). Basically, these materials form a sort of inert background, yet they possess an adequate number of groups to which selected functionalities can be added (Figure 2b). This kind of material represents valuable and versatile tools whose chemical/physical features can be independently modulated to a large extent. Natural fibrillar hydrogels, like collagen, fibrin or gelatin hydrogels, are endowed with ligands to which cells can adhere. Conversely, synthetic and polysaccharide hydrogels have to be modified in order to correctly display binding sites. This can be achieved by conjugating short peptide sequences or biomacromolecules (collagen, fibronectin) to the polymer backbone.

While the mechanisms on cell adhesion in 2D setups have been extensively characterized, the composition and dynamics of cell binding in a 3D environment are less defined. In fibrillar collagen gels, proteins, such as vinculin, paxillin, zyxin and talin, were found [44,45]. Additionally, while FA length correlates with substrate stiffness in 2D, long FAs are observed in 3D soft matrices provided that fibrils are coaligned with the FA axis. Furthermore, the level of zyxin and vinculin correlated with FA size [46]. Taken together, these data depict an intricate scenario in which adhesion dynamics, composition and morphology in 3D are affected by multiple chemical/physical features of the microenvironment, and the extrapolation of 2D results in a 3D context is not straightforward. This notwithstanding, there is growing evidence that adhesion molecules play an important role in 3D mechanotransduction in a similar manner as in 2D setups.

## 3. Engineering Materials to Control Cell Fate and Functions

### 3.1. Surface Patterning

Cells adhere to surfaces through a limited and well-defined number of points, *i.e.*, focal adhesions. Even in the case of surfaces uniformly coated with ligands, adhesion occurs in discrete locations. In fact, the maturation of adhesion involves the recruitment of transmembrane and cytoplasmic molecules at the site of adhesion. If this process is halted, for instance by too few or sparse ligands, adhesions can disassemble. Therefore, the presence/absence of the ligand is not the only parameter that affects adhesion events, but it is rather the way the ligand is presented that plays a non-negligible role. For instance, ligands' mobility (through flexible tethers), their density, spatial positioning, along with substrate stiffness and topography are all factors that eventually influence FA establishment, maturation and activation of signaling pathways consequently. Evidence of this was provided by Maheshwari *et al.*, who engineered PEG star-based polymeric substrates displaying YGRGD ligands in a clustered form, *i.e.*, 1, 5 or 9 ligands per star. Additionally, the authors were able to control the intercluster spacing that varied from 6–300 nm and achieved RGD surface densities in the $0.9 \times 10^3$–$1.2 \times 10^5$ ligand/cm$^2$ range [47]. NR6 fibroblasts displayed increased spreading and speed when the ligand was presented in a clustered form with respect to what was observed on non-clustered arrangements. Interestingly, a minimum cluster distance of 60 nm was required to permit adhesion and stress fiber formation, whereas 6 nm was calculated to be the threshold value for non-clustered ligands. These data point out that both ligand clustering and interligand spacing are important parameters in affecting cell adhesion.

The high sensitivity cells possess in recognizing ligand arrangements is not limited to differences in density or clustering. For instance, gradients of signals provide cells directional information necessary for polarization and migration, and various gradients of either bound or soluble signals are found *in vivo* [48,49].

A broad spectrum of methods was developed to generate concentration gradients of ligands on synthetic substrates. Methods based on plasma or light irradiation, diffusion, microcontact printing (μCP) and microfluidic, reviewed in Wu *et al.* [50], proved to be effective in generating gradients of ligands and enabled a precise control on gradient slope and average concentration. Combining photochemical and electrochemical approaches, Lee *et al.* fabricated RGD gradients on electroresponsive SAMs [51]. The authors studied the migratory response of 3T3 fibroblasts on different gradient slopes. Fibroblasts were very sensitive to both local ligand density and slope. In fact, cells on steep gradients terminated their migration in regions with a higher local RGD concentration with respect to cells migrating on shallow gradients. Furthermore, the authors showed the importance of FAK in sensing ligand presentation, as knockout FAK cells positioned themselves to the same density irrespective of the gradient slope. Concerning migration speed, Smith *et al.* used a diffusion-based method to realize fibronectin gradients on SAMs [52]. Endothelial cells showed a drift speed that correlated with gradient slope, whereas the random component of speed, along with the persistence time remained constant. Possibly, this behavior may arise from higher frequencies of cell polarization or its increased stability at higher gradients. Analogous results were obtained by Guarnieri *et al.*, who analyzed NIH/3T3 migration atop PEG hydrogels with RGD gradients fabricated with a fluidic gradient generator [53]. The authors found increased cell alignment on steep gradients. Drift speed correlated with gradient slope up to a threshold value for very steep gradients. These data, along with others, clearly demonstrated that specifically-engineered platforms displaying continuous gradients of ligands in biologically-relevant concentrations represent a valuable tool not only to unravel the basic mechanism underlying cell response to signals in a density-dependent manner, but also to effectively define the optimal concentration of bound signals for a specific cell response [54].

The methods described above, while instrumental in shaping our understanding of signal recognition and cell response, are not able to provide an accurate control of ligand spatial positioning on a micro- to nano-metric scale. This is a central aspect to create perfectly controlled environments for performing systematic studies on cell adhesion events. The implementation of micro- and nano-fabrication technologies allowed functionalizing material surfaces with a high spatial resolution and with reasonable costs and processing times. Photolithographic techniques are probably the cornerstone of all of the surface functionalization technologies aimed at fabricating functional surfaces to control cell adhesion. Photolithography consists of the exposure of a substrate, typically silicon, coated with light-sensitive photoresist with a patterned UV light. Light patterning is most conveniently performed by applying a specifically-designed reflective mask. In case of 'positive' photoresists, only those parts exposed to the radiation are soluble in organic solvents. Conversely, in 'negative' photoresists, the solvent dissolves the non-exposed parts, thus creating an inverse pattern. This process requires specialized equipment and high capital costs. However, the technology is nowadays very well consolidated, and raw materials are easily affordable: therefore, the fabrication of patterned surfaces can be outsourced. This allowed the development of soft lithographic techniques in which an elastomeric stamp or master, usually in PDMS, is fabricated and employed to create patterned surfaces. Patterned stamps can be treated with oxygen plasma to improve hydrophilicity and/or can be coated with proteins to promote cell adhesion and then can be used directly as substrates for cell cultures.

Thus defined, soft lithographic techniques encompass a broad spectrum of processes, among which replica molding (REM), µCP and micromolding in capillaries (MIMIC) have been extensively used to confine cell adhesion sites with a micrometric or sub-micrometric spatial resolution [55,56]. REM of synthetic polymers consists of embossing the topographic features of the elastomeric stamp onto polymer precursors or melted polymers, which are then solidified. Patterned elastomeric masters or structures fabricated via REM have been largely used to assess the effects of topographic features on cell behavior. Systematic studies on this issues regarded patterns in the form of gratings, pillars and protrusions. Basically, these substrates display 'terraces' on which cells can form adhesions, juxtaposed to recesses that might not be readily accessible. Topographic patterns having too narrow and/or too deep features might not allow the cell membrane to accommodate surface contours, thus causing the cells to be 'suspended' on the top of ridges and pillars. Furthermore, if these structures have lateral sizes, which may interfere with the normal formation and maturation of FAs, then alterations in cell adhesion, spreading orientation and migration are observed. In the case of parallel nanogratings, FAs and stress fibers are predominantly oriented along the pattern direction. In this circumstance, most of the cell types exhibit an elongated morphology and migrate preferentially parallel to the pattern direction [57]. This phenomenon, usually referred to as contact guidance, ceases to exist when topographic features are so shallow to not be recognized by the cell anymore. Apparently, the threshold depth below which features do not exert their regulatory role on migration and alignment is 35 nm [58].

To gain a better insight into the origin and effect of contact guidance on nanopatterned substrates, we used fluorescent tags to investigate the dynamics of FAs and cytoskeleton assemblies [59]. We found that actin fibers with directions different from that of the pattern possessed dashed adhesions that colocalized in the proximity of consecutive ridges. Such a peculiar assembly was caused by the confining effect induced by the nanopattern on FA growth. FAs thus formed were unstable and rapidly collapsed under the effect of actin-generated forces. Eventually, the vast majority of FAs were coaligned with the pattern direction, which in turn affected cytoskeletal structure and cell shape.

Micron- and submicron-scale patterns were also shown to affect cell proliferation [60,61]. However, literature studies are not conclusive on this aspect, as it seems that no obvious trends exist that allow one to predict the effects of topographic patterns on proliferation [62].

μCP uses the elastomeric master as a stamp to transfer molecules or proteins on surfaces. Usually, the surfaces are composed of or coated by a cell-repellent material, for example a self-assembled monolayer (SAM) of PEG. In this case, a sharp mismatch in adhesion properties is made, which results in the confinement of cell adhesion. Csucs *et al.* transferred patterns, with lateral resolution down to 1 μm, of adhesive molecules (either peptides or proteins) on various materials [63]. Adhesion mismatch was induced by poly-l-lysine-g-PEG backfill. This work demonstrated that through a careful optimization of the material properties and patterning procedure, the pattern was made very stable, even in the presence of serum proteins, which might in principle alter the ligand distribution on the surface. In fact, cells adhered on the functionalized regions only, and a strong directional confinement was observed during cell migration. More recently, Eichinger *et al.* proposed the development of the conventional μCP technique for multi-molecule transfer [64]. The development involves the use of modified inverted microscopes for proper stamp alignment prior to printing. The authors fabricated alternating micro-stripes of laminin and aggrecan and showed that astrocytes correctly recognized the multi-molecular pattern and adhered onto the laminin stripes only. This example extends the range of potential applications of μCP in settings requiring complex multimolecular patterns.

In MIMIC, a patterned elastomeric stamp with an open network of channels is pressed against the surface that needs to be functionalized. A solution containing the 'functionalizing' molecule is delivered through the network by capillary suction. The solution can be composed of polymer precursors or proteins. Solutes in the fluid can then be adsorbed on one surface or can be treated chemically or thermally, thus replicating the pattern features of the network. This method proved to be straightforward and effective in confining cell adhesion at a single [65] or multiple cell level [66].

With the above-mentioned methods, the size of pattern features displayed by the elastomeric stamp is limited by the diffraction of the UV light. Submicron-scale features can be obtained by using, for example, extreme UV light. Methods aimed at challenging the diffraction limit of light, such as electron beam lithography (EBL) and focused ion beam lithography (FIB), were developed in order to fabricate nanoscale features. These techniques use short wavelength electromagnetic sources and do not require a mask, as the beam is deflected on the surface with electromagnetic lenses. Additionally, FIB allows atoms to be displaced from or deposited onto the material surface, in which case it is possible to achieve subtractive or additive lithography on the final substrate directly, without further development. These techniques acquire particular importance when a spatial control on single integrin clusters or even individual ligands are required.

Using EBL combined with imprinting lithography, Schvartzman *et al.* fabricated arrays of metallic clusters constituted by AuPd nanodots assembled in different arrangements from dimmers up to heptamers. Each dot displayed a single RGD ligand, thus exerting a remarkable control on adhesion events at a single molecule level. The authors found that the overall density of dots did not affect spreading dramatically, whereas cluster size was crucial. Furthermore, tetrameric clusters of ligands, with interligand spacing of 60 nm, were necessary to enable cell spreading [67].

The fabrication of patterned surfaces exhibiting extremely small features is achieved at the expense of the processing time, which may render these nanotechnologies not particularly suitable when large-area patterning is required. A partial solution to this issue is represented by micellar lithography that involves the spontaneous arrangement of polymeric micelles, with a nanometric metal core, in a closely-packed quasi-hexagonal lattice on material surfaces [68]. For a careful modulation of the processing conditions and micelle characteristics, patterns with spacing ranging from 28–85 nm were reported. Metallic nanoparticles can be decorated with ligands or other biomolecules; therefore, by controlling the particle size and interparticle spacing, a tight control over FA formation and maturation can be achieved. By exploiting this technique, Arnold *et al.* fabricated quasi-hexagonal patterns of RGD functionalized gold nanodots with ligand spacings of 28, 58, 73 or 85 nm [69]. The authors observed a sharp transition of cell response in terms of adhesion and spreading passing from 58–73 nm. In particular, ligand spacing above 73 nm did not favor FA formation and actin assembly. Therefore, a ligand spacing of 58 nm is necessary to permit integrin clustering, thus triggering the cascade of events that lead to adhesion formation. Follow-up studies involving substrates displaying gradients of nanodot spacing (50–250 nm) found that a spacing slope of 15 nm/mm is the minimal slope required to induce cell polarization and demonstrated the exquisite sensitivity of cells in recognizing small spatial variations in ligand separation (~1 nm) [70]. These data, together with those of Maheshwari and Schvartzman, indicate that ~60 nm is a characteristic ligand distance above which adhesion formation is impaired. It was suggested that such a distance is required for talin binding, which then stabilizes integrin clustering [67]. While these figures seem to be consistent among various anchorage-dependent cell types, maximum interligand spacing of 32 nm was shown to be necessary for hematopoietic stem cells to adhere [71]. In fact, above such a threshold value, integrin clustering and lipid raft-dependent integrin signal transduction were strongly depressed.

### 3.2. Hydrogel Engineering

A large body of literature has been produced in the past few decades on hydrogel engineering to specifically control cell functions. These include cell adhesion, migration, spreading and differentiation. To this aim, hydrogels must be endowed with specific biological activities and microstructural features. First, cells have to be encapsulated within the hydrogel. Second, whatever chemical functionalization is chosen, this should not harm cells during the encapsulation process nor in culture. Concerning encapsulation, cells are usually suspended in a prepolymer solution. Gelation can be promoted by the insertion of reactive crosslinking molecules or it might rely upon physical forces. Various chemical crosslinking schemes have been developed, which proved to be cytocompatible and do not affect hydrogel bioactivity. These schemes include radical polymerization, click reactions and Schiff base crosslinking, reviewed in [72,73]. Usually, to gain a better spatial and temporal control on the reaction process, light-sensitive molecules are employed, which generate radicals upon exposure to the adequate radiation. Various combinations of polymers-photo initiators have been proposed so far [74]. However, some concerns have been raised on the toxicity of the photo-initiator and the UV radiation, which prompted the development of cell-friendly photo-crosslinkers [75].

Natural fibrillar hydrogels intrinsically possess peptide sequences that promote cell adhesion; they do not possess the adequate versatility in tuning the biochemical, structural and mechanical features independently, which is a fundamental requisite to exert a tight control on cell adhesion and response. Conversely, synthetic and polysaccharide hydrogels are characterized by chemical structures that make them more prone to achieving an orthogonal control on their biochemical/biophysical properties.

Different strategies can be pursued to engineer the biochemical/biophysical properties of hydrogels. These can be roughly grouped into top-down or bottom-up approaches. The former consists of chemical strategies that are optimized to conjugate active molecules or sequences to structural, possibly inert backbones, mainly of synthetic origin, thus creating a hybrid system. By exploiting this approach, Lutolf *et al.* formulated a PEG-based hydrogel containing both RGD- and MMP-sensitive domains [76]. This allows an orthogonal control on cell adhesion and cell-mediated degradation, which are elements of paramount importance to systematically investigate cell behavior *in vitro* or *in vivo*. Fibroblasts were able to break down the matrix by expressing MMP. Furthermore, their response, in terms of invasion, to ligand concentration was similar to what was observed in 2D setups, with cells displaying a biphasic dependence of migration on ligand density [77]. Finally, the authors proved the effectiveness of the hydrogel system in promoting bone regeneration in a critical size defect model. Taken together, this landmark study demonstrated that specifically-engineered hydrogels possess an enormous flexibility and versatility in the design of the material properties. This allows developing not only valuable tools to investigate complex cell-matrix interactions, but also systems that exert a therapeutic potential *in vivo*.

Hybrid systems were also designed in order to exploit peptide crosslinks, which form in mild conditions. For instance, Sanborn *et al.* fabricated a four-arm PEG functionalized with a fibrin-mimetic peptide that crosslinks in the presence of calcium divalent ions, thrombin and factor XIII, thus generating an elastic gel at 37 °C [78]. More recently, Ehrbar *et al.* followed a similar approach by functionalizing PEG macromers with specific substrate peptides that induced gel formation in the presence of Ca++ and factor XIII [79]. The use of peptide-synthetic hybrids is beneficial not only for modulating the hydrogel features, but it also allows encapsulated cells to perform complex functions. In the example above, human fibroblasts were able to cleave the peptide sequences of the network through matrix metalloproteinases (MMPs), which ultimately resulted in network remodelling with the gel forming extensive dendrite-like connections.

Many other biologically-derived features were implemented in synthetic hydrogels. For instance, the natural mechanism of growth factor (GF) sequestration/release via ECM binding molecules was implemented in synthetic hydrogels by means of different techniques [80,81]. On-demand release of active GF was achieved by Zisch *et al.*, who developed a PEG-based hydrogel containing both RGD for cell adhesion and VEGF coupled with MMP-sensitive domains [82]. This system has the advantage of avoiding systemic release of factors and limits its activity only upon cell-mediated proteolytic remodelling. This and other examples discussed in the following section demonstrate once again the superior capabilities of synthetic hydrogels in controlling the development of complex biological processes thanks to their ability to enable an orthogonal modulation of mechanical, biochemical and structural properties.

Top-down approaches, while beneficial for the realization of 3D macroscopic systems with clearly-defined biological properties, might not allow for accurate spatial arrangement of the bioactive, functional or structural elements. If spatial control of signals' display is desired, bottom-up approaches are more adequate. For instance, promoting or disrupting crosslinks in specific locations with patterns of light proved to be effective in creating hydrogels with well-defined microstructural features down to the sub-micrometric range.

Patterning of biochemical, mechanical, topographic signals can be achieved in 2D with consolidated and widespread techniques. Hydrogels, or more generally, 3D patterning, pose non-trivial technical hurdles. Patterned polymeric stamps have been used to confine hydrogel shape during gelation, thus creating simple networks of channels. By using such a method, Nelson *et al.* generated branched micro-channels in collagen gels and reported that their geometry can control epithelial morphogenesis by dictating the local microenvironment [83]. This method, although simple and robust, cannot be employed for the fabrication of complex 3D architectures. This notwithstanding, several processes have been developed that are able to locally manipulate gel structure or chemistry in a consistent and effective manner. In this part, we will present a limited number of hydrogel patterning technologies that were effective in spatially arranging signals, thus affecting cell behavior. Comprehensive reviews on 3D pattering technologies can be found in the specialized literature [84,85]. Basically, the technologies can be divided into two macro-sectors: spatial control of the gel crosslinking and spatially-controlled deposition of materials (that usually gels immediately after deposition on a supporting material or structure). The first category is dominated by light-induced crosslinking or dissolution. Stereolithographic methods rely on focusing of a light beam in a bath containing light-sensitive molecules that enable gel formation upon exposure to light. To produce 3D macroscopic objects, the fabrication occurs in a layer-by-layer fashion. Different approaches can be pursued to achieve the desired 3D structure with a 1–10-μm spatial resolution. In one approach, a fabrication platform is located beneath the prepolymer liquid interface and moves downwards as the hydrogel layers are inscribed on the platform top. Alternatively, the fabrication starts with the platform close to the prepolymer bath bottom (made of a transparent glass). As the hydrogel layer is formed by the light beam, the fabrication platform moves upwards, carrying along the newly-formed hydrogel structure. Platform movement is controlled by motors, which must allow for the micrometric motions, and the step between two adjacent layers is smaller than the curing depth produced by the light beam. The light source is generally a laser whose beam path is controlled by micro-tilting mirrors or stages. More recently, digital mirror devices or LCD screens have been employed to project light patterns on or beneath the prepolymer bath [86,87].

These techniques have been used to fabricate hydrogels with predefined ordered structures by employing either synthetic (poly(2-hydroxyethyl methacrylate) or PEG-diacrylate [88,89]) or natural (gelatin or alginate [90,91]) materials.

As previously discussed, the spatial resolution of hydrogel features is limited by the diffraction of light. Ten micrometer-wide features can be easily achieved. To improve spatial resolution, thus fabricating narrower features, different writing technologies have been developed. Along this line, multiphoton microscopy proved to be effective for hydrogel patterning. In fact, the focal point of femtosecond near-IR light is used to initiate the crosslinking reaction in a very small volume, leaving the rest of the prepolymer unaffected. By moving the focal point in the space, it is possible to fabricate complex structures with sub-micrometric resolution. Also for this technique, synthetic (PEG-diacrylate [92]) or natural (gelatin [93]) natural materials can be used.

Careful optimization of the processing conditions and crosslinking mechanism allows encapsulating cells within the patterned gel during fabrication, directly. For instance, Chan *et al.* investigated the viability of NIH-3T3 cells encapsulated in PEG hydrogels patterned with stereolithographic methods [94]. According to the processing conditions, the authors reported good cell viability and homogeneous seeding after seven days of culture. These data demonstrate the feasibility of patterning cell-hydrogel hybrids *in situ* with a high spatial resolution.

Material deposition-based approaches require the materials in the form of prepolymer to be extruded through a dye. Crosslinking occurs immediately after deposition on a supporting material with the gelation process being driven by external stimuli. For instance, Tirella *et al.* were able to fabricate honey-comb-like structures of alginate, crosslinked upon exposure to a $CaCl_2$ solution using a pressure-assisted micro-syringe technique [95]. The authors achieved hydrogel feature sizes down to 200 μm. A similar technology is represented by 3D-bioplotting; in this case, the prepolymer is extruded in a coagulation bath in which gelation might occur via temperature changes or chemical reactions induced by the medium. Generally, 3D-bioplotting allows the gelation to occur in mild conditions, which is amenable for direct cell encapsulation [96].

The fabrication of a fine network of hollow channels was made possible by subtractive methods. Recently, Miller *et al.* exploited a plotting-based technique to create an ordered network of micron-scale capillaries within a collagen gel [97]. Briefly, carbohydrate glass fibers, with diameters down to 200 μm, were plotted in a regular 3D lattice. Hydrogel precursors (either agarose, alginate, fibrin, Matrigel or PEG) were poured onto the network, allowing the solution to fill the porosity. Dissolution of the fibers generated a hollow and interconnected regular porous network. The authors fabricated a perfusable network of micro-vessels, lined with endothelial cells, surrounded by a cell containing gel, thus generating a vascularized tissue mimic *in vitro*. Subtractive patterning was also achieved by exploiting thermal degradation of materials. Using a near-infrared femtosecond laser, Hribar *et al.* controlled the thermal denaturation of cell-populated collagen-gold nanorod composites, thus creating channels with diameters down to ~8 μm, whilst maintaining good cell viability [98].

Spatial control of laser light was also employed to precisely locate biological functions of hydrogels, thus exerting a localized control of cell response. Kloxin *et al.* synthesized photodegradable PEG hydrogels' remote manipulation of gel properties *in situ* with UV irradiation of a two-photon laser [99]. Dissociation occurred upon irradiation, which resulted in the formation of channels that confined cell migration. The same concept was used to locally alter biochemical properties by conjugating bioactive moieties, here RGD, with photolabile sequences. Follow-up studies further demonstrated that the technology allows for a precise and predictable tuning of the gel structure in real time, which results in the control of the behavior of individual cells [100]. Lee *et al.* used laser light to generate patterns of RGD ligand within collagenase-sensitive PEG hydrogels [101]. RGD ligands covalently bound to the pre-gelled PEG network in the regions irradiated by the laser beam only. Human fibroblasts showed guided 3D migration only into the RGD-patterned regions of the hydrogels.

Sophisticated 3D-bioplotters equipped with multi-nozzle lines to deliver multiple material types have also been developed in order to fabricate organotypic cultures *in vitro*. Lee *et al.* assembled a pneumatic-driven four-channel plotter able to extrude collagen fibers or cells in a coagulation bath, thus creating a multi-layered skin equivalent constituted by dermis and epidermis strata [102]. Concerning lateral resolution, the definition of a limiting feature size depends on several parameters, such as dye shape, extrusion process, crosslinking method and material type. This stated, a broad range of lateral resolutions has been reported in the literature [103] spanning from a few up to hundreds of microns.

Recently, printing-based approaches have been extended to hydrogel materials. While the technology is more suitable for solid materials, specifically-designed apparatuses allowed the fabrication of patterned 3D structures with micron-scale resolution. Lam *et al.* printed powders of a polysaccharide gelatin blend that were bound together by water [104]. The use of water as a binding agent possesses the advantage of enabling material gelation in mild and cytocompatible conditions, in which the use of labile factors and biomolecules can be envisaged. An issue might arise on the mechanical strength of the final product; therefore, post-processing reinforcements are usually required. Similarly, Xu *et al.* fabricated a multi-layered fibrin hydrogel in the form of a fibrillar matt for *in vitro* neuron cultures [105]. The authors printed thrombin solution (crosslinking agent) over a layer of fibrinogen (fibrin precursor). Upon gelation, neurons were printed on the fibrillar hydrogel. The operation was reiterated five times, thus producing a 3D scaffold. Additionally, the authors reported the maintenance of phenotypic electrophysiological fingerprints for neurons seeded in 3D.

Advancements in biomaterial synthesis and plotting technologies allowed the development of bioprinting-based techniques in which "bioinks" made of cells and supporting materials are delivered in an orderly on a scaffolding "biopaper". The main component of the bioink is represented by cells, which are preliminarily assembled in the form of spheroids or cylinders. These are plotted on an inert hydrogel, the biopaper in the desired shape. Additional hydrogel struts might be required to generate complex structures, such as hollow cylinders. Then, the process is reiterated in a layer-by-layer fashion. Printed tissues/organs are cultivated in incubators, and the spheroids fuse together. The structure then matures, undergoing morphogenetic events reminiscent of early embryonic development. In this process, the presence of a hydrogel is crucial, as it provides a supporting frame that dictates the final shape of the construct. In particular, when different cell types are used to form the bioink particles, a segregation of cells occurs, forming cellular patterns observed in natural tissues. This phenomenon was exploited to generate complex tissues, such as blood vessels [106], cardiac tissue [107] and nerves [108].

It should be pointed out that many of the hydrogel patterning technologies described above were predominantly aimed at tissue engineering applications or at guiding the behavior of cell populations. Current limitations of the 3D patterning technologies do not allow exerting a fine control on the spatial positioning of ligands or topographic features at the nanoscale, which is necessary to control adhesion processes at a single cell level. However, the steady advancements in optical methods, for example two-photon microscopy, hold the promise to enable 3D hydrogel pattering with a nanometric resolution and in cytocompatible conditions.

## 4. Mechanosensing and Mechanotransduction

The functions of FAs are not limited to cell adhesion. Many signaling proteins, such as FAK, ERK, JNK and Src, are contained within FAs and are involved in a broad spectrum of signaling pathways that regulate diverse aspects of cell behavior, such as migration, proliferation and differentiation [109]. Therefore, besides their mere structural functions, FAs are important signaling centers. Since FAs' shape, composition and dynamics are very sensitive to exogenous stimuli, it is in principle possible to control complex cellular functions through the modulation of FA formation and growth. The examples described above clearly demonstrate that ligand presentation, in terms of spatial density and clustering, is an effective manner to modulate FA dynamics. Moreover, since many molecules participating in FA formation are mechanosensitive, change their activity and interact differently with other partners if subjected to actin-generated forces, the manipulation of cell contractility via material stiffness also affects adhesion-related signaling pathways and eventually cell functions. In fact, there is growing evidence that these approaches, *i.e.*, controlling ligand density, availability and positioning through biochemical or topographic patterns, or modulating FA dynamics with material stiffness, lead to strikingly similar effects on cell functions and fate.

Cell sensing of and response to material stiffness has been traditionally studied on synthetic hydrogels. For instance, PAM hydrogels were employed to study cell adhesion, migration and differentiation in experimental conditions in which stiffness could be varied in ranges similar to those measured in natural tissues. Our current understanding of mechanosensing and mechanotransduction is predominantly based on 2D setups, and yet, many biological mechanisms still need to be thoroughly elucidated. This notwithstanding, it is now widely accepted that adhesion formation and stress fiber contraction play a central role in mechanosensing [110]. In this process, the way ligands are anchored to the substrate, for instance either with flexible or rigid tethers, is crucial. More specifically, when cells pull the substrate, the actual stiffness they perceive is an integrated mechanical property in which both material elasticity and ligand mobility play a role. Concerning the latter aspect, Houseman *et al.* conjugated RGD on glass by means of oligo(ethylene glycol) spacers of various lengths [111]. The authors found that for a fixed density of peptide, increasing the length of the spacer significantly decreases the efficiency of cell attachment and spreading. Conversely, Kuhlman *et al.* developed a poly(methyl methacrylate)-graft-poly(ethylene oxide) system in which PEO segments of different lengths displayed RGD [112]. The authors found that the longer tethers increased the rate of fibroblast spreading and reduced the time for FA formation. Presumably, the tether flexibility favored integrin clustering and FA formation. This apparent discrepancy between the findings might arise from the differences in the range of tether lengths and ligand densities that were investigated. However, these data demonstrated the high sensitivity that cells possess in recognizing and reorganizing signals at the nanoscale. Along this line, Choi *et al.* have recently reported the development of a facile platform to study the influence of RGD ligand coupling strength on hMSCs' spreading and differentiation, while maintaining substrate hydrophilicity and ligand density constant [113]. The authors reported enhanced osteogenesis and Yes-associated protein activity on tightly-bound RGD substrates that confirm the importance of ligand elasticity in dictating stem cell fate.

This notwithstanding, the regulatory role of bound signal stretching on cell fate and functions has not been thoroughly understood. Trappmann *et al.* highlighted that ligand tethering can be influenced by substrate stiffness [114]. In fact, the authors found that stiffer PAM hydrogels possess a smaller mesh size that offers more anchoring points for bioadhesive molecules. This results in an enhanced mechanical feedback. More recently, Wen *et al.* used an analogous material platform, namely PA gels, to study MSC response to changes in substrate stiffness, porosity and ligand tethering [115]. Differently from the previous work, the authors demonstrated that the main regulator of MSC fate was the substrate bulk stiffness, whereas porosity and tethering had negligible effects on cell fate. Other platforms were developed and presented in the literature that allow tuning substrate stiffness and adhesivity independently. For instance, Fu *et al.* developed arrays of microposts in PDMS on top of which cells were suspended [116]. Post deflection was observed upon cell contraction. With this setting, the bending stiffness could be modulated by changing the post height, while keeping adhesive properties unchanged.

The transduction of mechanical signals into biological events gained attention when Engler *et al.* showed that stem cell lineage specification was driven by PAM elasticity, with MSCs differentiating into neurons, myoblasts or osteoblasts when the stiffness of the underlying matrix approached the stiffness of brain, muscle or bone, respectively [117]. Stiff gels induced a more contractile phenotype and promoted osteogenesis, whereas less spread and contractile cells differentiated into adipocytes. Other studies, not directly exploiting matrix stiffness, but aimed at manipulating cell contractility through cell shape, reported similar observations, *i.e.*, shapes promoting myosin-driven contractility-enhanced osteogenesis, whereas round shapes induced little cell contractility and promoted adipogenesis [118–120]. This viewpoint has been recently challenged by Wang *et al.*, who observed increased osteogenesis and adipogenesis on RGD-functionalized gold nanodot patterns, which in principle should depress cell spreading, thus favoring adipogenesis [121]. Even though the underlying mechanisms have not been thoroughly elucidated, the authors suggested that RGD nanospacing might be an inherent signal that regulates differentiation of stem cells beyond cell spreading. Since the structure and stiffness of natural tissues are the result of continuous remodelling and biosynthetic processes that take place both in morphogenesis and during tissue homeostasis, it is not surprising that cells might respond differently if subjected to time changing material stiffness with respect to static materials. This aspect was investigated by Young and Engler [122], who cultivated cardiomyocytes on HA-PEG hydrogel that was subjected to a crosslinking program aimed at recapitulating the stiffening of native cardiac ECM during morphogenesis. The authors found that increasing matrix stiffness according to biologically-inspired time programs improved cardiomyocyte differentiation.

Most of the works dealing with the influence of substrate stiffness on cell adhesion are based on studies performed on single material types, chiefly PAM, PEG and PDMS. Therefore, possible effects arising from the chemistry of the material rather than its intrinsic mechanical properties might not emerge clearly. Additionally, substrate materials are usually treated as linear elastic materials. However, soft hydrogels and elastomers, as those used in mechanobiology studies, are viscoelastic in nature. Thus, the hypothesis of linear elasticity can be misleading when drawing out general conclusions on the role of material mechanical properties on cell behavior, especially when time-dependent effects are taken into consideration. The role of substrate viscoelasticity in altering cell mechanotransduction was carefully addressed by Cameron *et al.* [123]. The authors cultivated human mesenchymal stem cells (MSCs) on different PAM hydrogels, having constant stiffness (elastic modulus), but varying levels of creep (loss moduli). A decreased size and maturity of FAs and an increased proliferation, spreading and differentiation towards multiple lineages, with a propensity towards myogenesis, were observed for cells on high-creep hydrogels. The authors suggested that the decrease in isometric tension (*i.e.*, contractility) is compensated by an increase in isotonic tension, which favors spreading and more dynamic FAs. Follow-up studies [124] provided a deeper insight into the molecular mechanisms regulating fate specification. The authors found that the creep-induced loss of cytoskeletal tension is compensated by an increase in isotonic tension with an increase of Rac1 activity, which eventually promotes myogenesis. Similar concepts have been recently developed by Chaudhuri *et al.*, who cultivated U2OS cells on either purely elastic or viscoelastic alginate hydrogels [125]. In the range of low modulus hydrogels, increased spreading and stress fiber formations were observed on stress-relaxing substrates compared to purely elastic ones. Thus, the mechanism of stress relaxation compensates for the decreased stiffness. Possibly, ligand mobility enables integrin clustering that may enhance cell spreading [112]. Taken together, these data challenge the hypothesis that mechanotransduction is solely governed by cell resistance to elastic forces, as dissipative effects can have a non-negligible role in affecting cell behavior.

It has to be pointed out that the examples above refer to cells cultured on hydrogels, thus representing a 2D experiment. Few characteristic traits, arising from the chemical-physical characteristics of the microenvironment, differentiate 3D mechanosensing from its 2D counterpart. Natural fibrillar gels, like collagen or fibrin gel, possess a very heterogeneous microstructure in which both long and short fibrils, along with straight and slack fibrils coexist. Therefore, a single cell might experience very different mechanical feedback according to the local network heterogeneity. For instance, individual collagen fibrils are characterized by a modulus in the range of hundreds of MPa under traction [126], whereas fibril bending or pulling buckled fibrils would produce very different mechanical feedbacks (Figure 3).

Along these lines, Kubow *et al.* showed that long adhesion formed along straight fibrils, whereas small adhesions were found on angled, retracting fibrils [46]. Additionally, they reported increased amounts of vinculin and zyxin in long 3D adhesions, which is a characteristic similar to that observed in 2D. In a different experimental setup, Paszek *et al.* [127] reported reduced FAK phosphorylation at Y397 within 3D collagen gels with respect to a stiff 2D glass surface and that phosphorylation levels correlated with gel stiffness. Extensive and prolonged fibril pulling might cause dramatic network remodelling, which often results in dramatic gel compaction (Figure 4). Unconstrained gels can be compacted up to 20% of their initial volume, whereas fibril alignment is observed when gel edges are blocked, causing compaction to occur in specific planes [128,129].

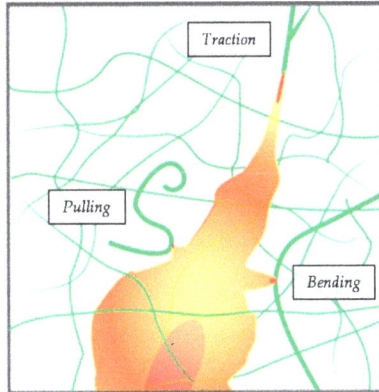

**Figure 3.** Cell perception of mechanical cues in 3D fibrillar gels. Cells might receive very different mechanical feedbacks from the surroundings according to the local configuration of the matrix. High traction forces are required for direct fibril traction, whereas low forces are necessary for fibril bending or pulling slack fibrils. Focal adhesion (FA) length and composition change accordingly.

(a)  (b)

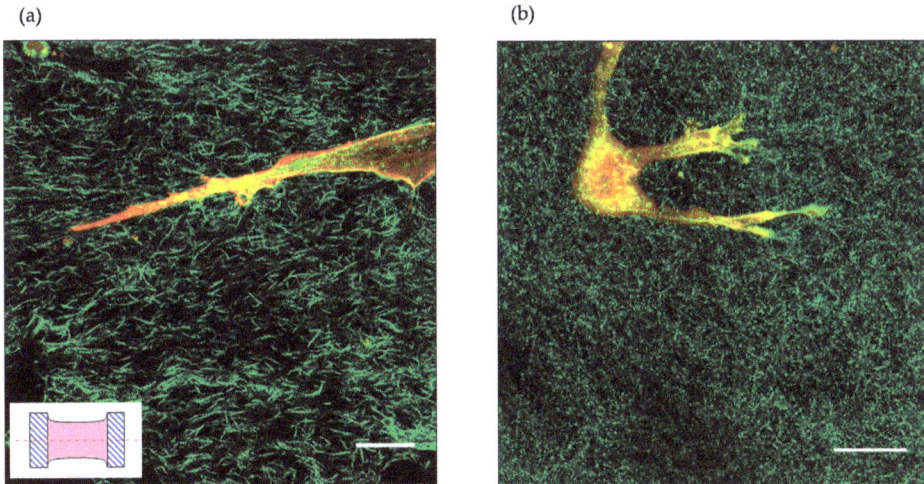

**Figure 4.** Cells encapsulated in type I collagen gel. Bovine dermal fibroblasts cultivated for one week in 2.4 mg/mL collagen gel. (**a**) In uniaxial constrained gels, cell-generated forces cause the gel to compact in the two free directions (orthogonal to the constraints; see the inset). Upon compaction, collagen fibrils result in being aligned and cells are elongated. (**b**) In free floating gels, extensive collagen densification is observed around cells, which possess a dendrite-like morphology. Collagen (green) is observed in reflection mode. Cells are stained for actin (TRITC-phalloidin). Bar = 20 μm.

This makes it hard to understand the mechanosensing mechanisms, since a local increase in fibril density, stiffness and reduced pore size occur during gel compaction. Taken together, these data suggest that while some aspects of 3D mechanosensing might find analogies with 2D setups, the intrinsic complexity of the 3D environment makes it difficult to draw out definitive conclusions on the molecular pathways governing cell recognition and reaction to mechanical stimuli, as these appear to be strongly influenced by hydrogel and cell type and experimental setups. This prompted the development of artificial systems in which stiffness and ligand availability can be varied independently. Among the various materials that have been proposed, PEG-based hydrogels have been extensively used to study mechanosensing and mechanotransduction in 3D. Other molecules, such as alginate, agarose or derivatives of hyaluronan, also offer a comparable degree of versatility in terms of manipulation of the hydrogel chemical/physical properties. The elegant study from Huebsch *et al.* clearly demonstrated how dimensionality alters the cell perception of the surrounding environment [130]. The authors encapsulated murine MSC in alginate hydrogels with stiffness varying from 2.5–110 kPa. Differently from what was observed in 2D systems on which gross cell morphological changes correlate with MSC differentiation, here, fate decisions were uncorrelated with cell morphology. Specific levels of matrix elasticity favored integrin clustering, ligand reorganization and, therefore, cytoskeletal tension, which correlated with osteogenesis. In this study, osteogenesis occurred in the case of intermediate hydrogel stiffness. Khetan *et al.* were able to induce radical crosslinking in HA-peptide hybrids that spatially confined 3D cell cultures by inhibiting degradation in specific locations [131]. This was achieved by the mutual arrangement of proteolytic crosslinks and non-degradable kinetic chains through UV exposure. Even if exposed to an osteogenic mechanical environment (~18 kPa), human MSC were unable to spread and differentiated towards an adipogenic lineage. Conversely, cells remodeled the proteolytic-sensitive domains, which allowed them to exert greater tension that favored osteogenesis despite the low stiffness of the surrounding environment. Follow-up studies also demonstrated how time changes of hydrogel stiffness could dramatically affect stem cell differentiation. Guvendiren *et al.* developed a methacrylate-functionalized HA gel that can undergo sequential crosslinking [132]. Stem cells were encapsulated in soft gels for selected time intervals, after which gels were stiffened by light irradiation. Early stiffening (*i.e.*, cells exposed to a stiff environment for a long timeframe) promoted osteogenesis, whereas late stiffening promoted adipogenesis. In a more recent study, Khetan *et al.*, using HA with dynamically tunable degradation properties, investigated the mechanics governing lineage specification arising from mechanical cues in 3D [133]. Their system elegantly demonstrated that gel remodelling is essential in driving cell differentiation: even in a 'morphologically' spread state, cell were not able to launch an osteogenic program, owing to the low traction forces generated in a restrictive, non-degradable matrix. These examples demonstrated that matrix stiffness alone is not sufficient to predict stem cell differentiation in 3D, but it is rather the dynamic variation of stiffness and degradation that plays a crucial role in affecting cell fate.

## 5. Supramolecular Materials: Mimicking the Natural Environment of Cells

The ideal hydrogel for biomedical applications should merge the advantages of both natural and synthetic hydrogels, *i.e.*, microarchitectures similar to the native ECM and the extraordinary tunability of synthetic molecules; at the same time, such a hydrogel should minimize potential toxic effects arising from functionalization, especially when this are carried out in the presence of cells. Specifically engineered peptides forming highly hydrated nanofibrillar structures represent a very interesting class of materials that reproduces some aspect of the architecture of native ECM and allows great versatility in the modulation of the biochemical and biophysical features of the structure.

Early works by Zhang [134] showed that certain peptide sequences, inspired by natural proteins, undergo self-assembly in aqueous solutions. The peptide sequences were constituted by repetitions of hydrophilic and hydrophobic amino acids. Hydrophobic and ionic interactions between adjacent molecules cause the peptide to stack and form β-sheet structures, eventually producing nanometer-sized fibrils. Since this pioneering discovery, several peptides displaying self-assembling properties were designed. In particular, by changing the repeating amino acid sequence, it is possible to modulate hydrogel formation [135] and supramolecular organization in the form of β-hairpin or α-helices [136,137]. For example, Ramachandran *et al.* reported that mixing peptides exhibiting opposite-charged groups formed stable hydrogels that were sensitive to pH, salt and mechanical shearing forces [138]. Owing to their amenable properties of gelling in mild and cytocompatible conditions, self-assembling peptides have been widely used in Tissue Engineering applications. For instance, two formulations of peptides constituted by the repetition of arginine-alanine-aspartate (known as RAD-I and RAD-II) were used as a scaffold for neuronal regeneration. Even if these systems were not specifically functionalized with bioactive neurotropic moieties, RAD-based self-assembling hydrogels promote neuron adhesion and synapse formation [139]. Further studies were aimed at providing the gels with specific biologic functions to control cell response. Along these lines, Kumada and Zhang [140] and Kumada *et al.* [141] conjugated fibronectin- (RGD) and laminin- (PDS) derived cell adhesion sequences to RADA16-I hydrogels and reported increased fibroblast proliferation, migration and collagen biosynthesis.

MMP-sensitive domains were also incorporated into the gel building blocks in order to promote cell-mediated hydrogel degradation. Galler *et al.* engineered a self-assembling peptide hydrogel whose fibrils displayed RGD adhesion ligands and MMP2 cleavable segments [142]. These functionalities improved cell spreading and migration.

Taken together, these data demonstrate the enormous flexibility in designing peptide building blocks, thus endowing the resulting self-assembling hydrogels with biochemical/biophysical properties defined *ab initio*. Furthermore, owing to the "modular" nature of the building block structure in which individual sequences fulfil specific tasks in the self-assembling process and dictate the chemical features of the resulting network, it is in principle possible to exert an orthogonal control in engineering the final properties of the hydrogel without interfering with the spontaneous self-assembly.

Stupp's laboratory synthesized peptide amphiphiles that self-assemble under specific conditions (pH, temperature, ionic strength). Hydrogels could be engineered in a bottom-up approach since their properties are strictly related to the structure and amount of the individual building blocks. These are constituted by four distinct regions, and the structure of each region profoundly affects the chemical/physical properties of the self-assembled hydrogel [143]. In particular, the hydrophobic sequence of the middle region bestows upon the peptide the tendency of self-assembling in the form of β-sheets. Furthermore, modifications in the structure also affect the macroscopic mechanical properties of the gel [144]. The amphiphilic nature of the peptides is provided by a hydrophobic tail and a hydrophilic peptide sequence at the other end. The latter allows good solubility and prevents self-assembly. If their charged amino acids are screened by salts or the pH of the medium is changed, self-assembly starts, eventually generating an entangled meshwork of nanofibers. Functional sequences, such as ligands or small bioactive signals, can be conjugated to the PA head, without any loss in the self-assembling capability of the system [145]. Using this strategy, specifically-designed PAs were synthesized to produce bioactive hydrogels to control cell fate and functions. Storrie *et al.* conjugated RGD sequences to PA in the form of linear, branched or cyclic geometries and studied DA 231 epithelial cell behavior *in vitro* [146]. Improved cell adhesion, spreading and migration were observed when ligands were assembled in branched, low packed structures protruding out of the fibers. This arises from an optimal presentation of adhesive signals in space. In a different setting, Silva *et al.* used the laminin-derived sequence IKVAV to control neural progenitor differentiation. IKVAV-functionalized PA gels promoted neural differentiation and suppressed astrocyte formation [147]. Interestingly, the authors also reported that IKVAV was more effective than laminin in controlling neurogenesis; this probably arises from a higher density of signals effectively presented to progenitor cells.

The above-mentioned examples demonstrated the ability of supramolecular materials to control various aspects of cellular behavior and highlighted the great potentialities these systems might have both *in vitro* and *in vivo*. However, these materials still lack some properties that would make them ideal systems to control cell behavior. First, the dynamic modulation of supramolecular material properties is still in its infancy, and only a limited number of examples of stimuli-responsive systems has been reported [148–150]. Second, the structures of the molecular building blocks might not make these materials directly compatible with the conventional techniques for micro- and nano-patterning. However, recent advancements in synthesizing and characterizing building blocks with enhanced functionalities and self-assembling properties will certainly help in making supramolecular materials relevant and complete platforms to control and study cell response in a chemically-/physically-controlled environment.

## 6. Future Perspectives and Conclusions

The use of engineered materials able to control and guide cell functions *in vivo* is still limited in clinical practice, in which more conservative treatments employing permanent prosthesis or resorbable devices still are the preferred choices in a large number of treatments. This notwithstanding, this review demonstrated that the studies focusing on fields of controlling cell, in particular stem cell, behavior with material surfaces or functional hydrogels are enormously active. In the future, the outcomes of these research lines might merge together to provide novel technical solutions for relevant biotechnological and clinical applications. For instance, the development of specifically-engineered materials able to exert a tight control on cell adhesion processes will be beneficial, not only to unravel the basic mechanisms governing cell reactions to exogenous stimuli, but also to fabricate complex systems or devices that control selected cell functions, like proliferation, differentiation and tissue regeneration, in a highly effective manner. Concerning *in vitro* models, the above outcomes will definitely aid the development of devices to study tissue morphogenesis and pathology progressions in a chemically-/physically-defined environment. The generation of *in vitro* models, such as organoids, although feasible, does not fully exploit the guiding effects provided by material signals. This results in substantial morphologic and functional differences between *in vitro* organoids and their counterparts found *in vivo* [151]. Yet, material features proved to profoundly affect cell self-organization and differentiation *in vitro* [152]. Similarly, the *in vivo* application of instructive scaffolds requires engineering materials displaying a set of stimuli that elicit specific cell differentiation programs and biosynthetic events in a deterministic manner. However, our current knowledge on possible relationships between material stimuli and biological responses is still in its infancy. This lack of understanding of the intricate cell-material interactions strongly limits the potential clinical applications of engineered material systems. Generally, cells integrate a combination of stimuli, also different in nature, simultaneously. This makes it difficult to deconstruct the problem, and thus, assembling combinations of stimuli to affect cell fate in a predictive manner is arduous. Owing to these difficulties, different approaches are being pursued. Combinatorial approaches, based on the fabrication of libraries of materials displaying combinations of stimuli, have been developed [153–155]. Such high-throughput libraries, combined with advanced investigative techniques, might help in solving the problem, thus allowing engineering systems in a deductive manner for a deterministic control of cell fate and functions.

Furthermore, because cells are naturally exposed to signals that vary in space and time according to specific programs, materials systems that enable a dynamic control of their chemical/physical properties can in principle be more effective in controlling cell functions. Few examples in which changes in properties are tailored in order to improve cell response have been recently reported in the literature [156,157].

We envision that by combining material patterning at the micro- and nano-scale and the dynamic presentation of signals with the highly tunable mechanical and biochemical properties of hydrogels will make it possible to replicate the complex environment in which cells can fulfil specific tasks defined *ab initio*. Such engineered environments will certainly impact diverse fields, both related to *in vitro* and *in vivo* investigations, such as instructive scaffolds that guide stem cell differentiation and tissue genesis for tissue engineering and regenerative medicine, the development of *in vitro* models, such as organoids or tissue analogues for drug screening and discovery, or as developmental/pathological models in an *in vitro* context.

**Acknowledgments:** The author thanks Valentina La Tilla and Carlo F. Natale for providing original material for the figure panels' assembly. We thank Lucia Formisano and Maria Iannone for the helpful discussions and Roberta Infranca for the proofreading process.

**Author Contributions:** Maurizio Ventre gathered the literature material, wrote the review and assembled the figure panels. Paolo A. Netti organized the topics, contributed to the discussion of the contents and corrected the manuscript.

**Conflicts of Interest:** The authors declare no conflict of interest.

## References

1. Murphy, W.L.; McDevitt, T.C.; Engler, A.J. Materials as stem cell regulators. *Nat. Mater.* **2014**, *13*, 547–557. [CrossRef] [PubMed]

2. Hao, J.; Zhang, Y.; Jing, D.; Shen, Y.; Tang, G.; Huang, S.; Zhao, Z. Mechanobiology of mesenchymal stem cells: Perspective into mechanical induction of MSC fate. *Acta Biomater.* **2015**, *20*, 1–9. [CrossRef] [PubMed]

3. Ventre, M.; Causa, F.; Netti, P.A. Determinants of cell-material crosstalk at the interface: Towards engineering of cell instructive materials. *J. R. Soc. Interface.* **2012**, *9*, 2017–2032. [CrossRef] [PubMed]

4. Ventre, M.; Netti, P.A. Engineering Cell Instructive Materials To Control Cell Fate and Functions through Material Cues and Surface Patterning. *ACS Appl. Mater. Interfaces.* **2016**. [CrossRef] [PubMed]

5. Dupont, S.; Morsut, L.; Aragona, M.; Enzo, E.; Giulitti, S.; Cordenonsi, M.; Zanconato, F.; Le Digabel, J.; Forcato, M.; Bicciato, S.; *et al.* Role of YAP/TAZ in mechanotransduction. *Nature* **2011**, *474*, 179–183. [CrossRef] [PubMed]

6. Hamamura, K.; Swarnkar, G.; Tanjung, N.; Cho, E.; Li, J.; Na, S.; Yokota, H. RhoA-mediated signaling in mechanotransduction of osteoblasts. *Connect. Tissue Res.* **2012**, *53*, 398–406. [CrossRef] [PubMed]

7. Jain, N.; Iyer, K.V.; Kumar, A.; Shivashankar, G.V. Cell geometric constraints induce modular gene-expression patterns via redistribution of HDAC3 regulated by actomyosin contractility. *Proc. Natl. Acad. Sci. USA* **2013**, *110*, 11349–11354. [CrossRef] [PubMed]

8. Gattazzo, F.; Urciuolo, A.; Bonaldo, P. Extracellular matrix: A dynamic microenvironment for stem cell niche. *Biochim. Biophys. Acta* **2014**, *1840*, 2506–2519. [CrossRef] [PubMed]

9. Chen, S.; Lewallen, M.; Xie, T. Adhesion in the stem cell niche: Biological roles and regulation. *Development* **2013**, *140*, 255–265. [CrossRef] [PubMed]

10. Guvendiren, M.; Burdick, J.A. Engineering synthetic hydrogel microenvironments to instruct stem cells. *Curr. Opin. Biotechnol.* **2013**, *24*, 841–846. [CrossRef] [PubMed]

11. Provenzano, P.P.; Keely, P.J. Mechanical signaling through the cytoskeleton regulates cell proliferation by coordinated focal adhesion and Rho GTPase signaling. *J. Cell Sci.* **2011**, *124*, 1195–1205. [CrossRef] [PubMed]

12. Jansen, K.A.; Donato, D.M.; Balcioglu, H.E.; Schmidt, T.; Danen, E.H.; Koenderink, G.H. A guide to mechanobiology: Where biology and physics meet. *Biochim. Biophys. Acta* **2015**, *1853*, 3043–3052. [CrossRef] [PubMed]

13. Kanchanawong, P.; Shtengel, G.; Pasapera, A.M.; Ramko, E.B.; Davidson, M.W.; Hess, H.F.; Waterman, C.M. Nanoscale architecture of integrin-based cell adhesions. *Nature* **2010**, *468*, 580–584. [CrossRef] [PubMed]

14. Tamada, M.; Sheetz, M.P.; Sawada, Y. Activation of a signaling cascade by cytoskeleton stretch. *Dev. Cell* **2004**, *7*, 709–718. [CrossRef] [PubMed]

15. Sawada, Y.; Tamada, M.; Dubin-Thaler, B.J.; Cherniavskaya, O.; Sakai, R.; Tanaka, S.; Sheetz, M.P. Force sensing by mechanical extension of the Src family kinase substrate p130Cas. *Cell* **2006**, *127*, 1015–1026. [CrossRef] [PubMed]

16. Parsons, J.T.; Horwitz, A.R.; Schwartz, M.A. Cell adhesion: Integrating cytoskeletal dynamics and cellular tension. *Nat. Rev. Mol. Cell. Biol.* **2010**, *11*, 633–643. [CrossRef] [PubMed]

17. Cavalcanti-Adam, E.A.; Aydin, D.; Hirschfeld-Warneken, V.C.; Spatz, J.P. Cell adhesion and response to synthetic nanopatterned environments by steering receptor clustering and spatial location. *HFSP J.* **2008**, *2*, 276–285. [CrossRef] [PubMed]

18. Vladkova, T.G. Surface Engineered Polymeric Biomaterials with Improved Biocontact Properties. *Int. J. Poly. Sci.* **2010**, *2010*. [CrossRef]

19. Massia, S.P.; Hubbell, J.A. An RGD spacing of 440 nm is sufficient for integrin αVβ3-mediated fibroblast spreading and 140 nm for focal contact and stress fiber formation. *J. Cell Biol.* **1991**, *114*, 1089–1100. [CrossRef] [PubMed]

20. Truskey, G.A.; Proulx, T.L. Relationship between 3T3 cell spreading and the strength of adhesion on glass and silane surfaces. *Biomaterials* **1993**, *14*, 243–254. [CrossRef]

21. Grzesiak, J.J.; Pierschbacher, M.D.; Amodeo, M.F.; Malaney, T.I.; Glass, J.R. Enhancement of cell interactions with collagen/glycosaminoglycan matrices by RGD derivatization. *Biomaterials* **1997**, *18*, 1625–1632. [CrossRef]

22. Fusco, S.; Panzetta, V.; Embrione, V.; Netti, P.A. Crosstalk between focal adhesions and material mechanical properties governs cell mechanics and functions. *Acta Biomater.* **2015**, *23*, 63–71. [CrossRef] [PubMed]

23. Lord, M.S.; Modin, C.; Foss, M.; Duch, M.; Simmons, A.; Pedersen, F.S.; Besenbacher, F.; Milthorpe, B.K. Extracellular matrix remodelling during cell adhesion monitored by the quartz crystal microbalance. *Biomaterials* **2008**, *29*, 2581–2587. [CrossRef] [PubMed]

24. Llopis-Hernández, V.; Rico, P.; Ballester-Beltrán, J.; Moratal, D.; Salmerón-Sánchez, M. Role of surface chemistry in protein remodeling at the cell-material interface. *PLoS ONE* **2011**, *6*, e19610. [CrossRef] [PubMed]

25. Underwood, P.A.; Steele, J.G.; Dalton, B.A. Effects of polystyrene surface chemistry on the biological activity of solid phase fibronectin and vitronectin, analysed with monoclonal antibodies. *J. Cell. Sci.* **1993**, *104*, 793–803. [PubMed]

26. Uquillas Paredes, J.A.; Polini, A.; Chrzanowski, W. *Protein-based Biointerfaces to Control Stem Cell Differentiation, in Biointerfaces: Where Material Meets Biology*; Hutmacher, D., Chrzanowski, W., Eds.; Royal Society of Chemistry: Cambridge, UK, 2014; pp. 1–29.

27. Chaudhuri, T.; Rehfeldt, F.; Sweeney, H.L.; Discher, D.E. Preparation of collagen-coated gels that maximize *in vitro* myogenesis of stem cells by matching the lateral elasticity of *in vivo* muscle. *Methods Mol. Biol.* **2010**, *621*, 185–202. [CrossRef] [PubMed]

28. Anamelechi, C.C.; Clermont, E.E.; Brown, M.A.; Truskey, G.A.; Reichert, W.M. Streptavidin binding and endothelial cell adhesion to biotinylated fibronectin. *Langmuir* **2007**, *23*, 12583–12588. [CrossRef] [PubMed]

29. Pan, J.F.; Liu, N.H.; Shu, L.Y.; Sun, H. Application of avidin-biotin technology to improve cell adhesion on nanofibrous matrices. *J. Nanobiotechnol.* **2015**, *13*, 37. [CrossRef] [PubMed]

30. Masters, K.S.; Anseth, K.S. Cell- Material Interactions. In *Advances in Chemical Engineering Molecular and Cellular Foundations of Biomaterials*; Peppas, N.A., Sefton, M.V., Eds.; Academic Press: Amsterdam, The Netherlands, 2004; Volume 29, pp. 7–46.

31. Cosson, S.; Otte, E.A.; Hezaveh, H.; Cooper-White, J.J. Concise review: Tailoring bioengineered scaffolds for stem cell applications in tissue engineering and regenerative medicine. *Stem Cells Transl. Med.* **2015**, *4*, 156–164. [CrossRef] [PubMed]

32. Hersel, U.; Dahmen, C.; Kessler, H. RGD modified polymers: Biomaterials for stimulated cell adhesion and beyond. *Biomaterials* **2003**, *24*, 4385–4415. [CrossRef]

33. Elbert, D.L.; Hubbell, J.A. Conjugate addition reactions combined with free-radical cross-linking for the design of materials for tissue engineering. *Biomacromolecules* **2001**, *2*, 430–441. [CrossRef] [PubMed]

34. Campbell, I.D.; Humphries, M.J. Integrin structure, activation, and interactions. *Cold Spring Harb. Perspect. Biol.* **2011**, *3*, a004994. [CrossRef] [PubMed]

35. Ikada, Y. Challenges in tissue engineering. *J. R. Soc. Interface* **2006**, *3*, 589–601. [CrossRef] [PubMed]

36. Worley, K.; Certo, A.; Wan, Q.L. Geometry–Force Control of Stem Cell Fate. *BioNanoSci* **2013**, *3*, 43–51. [CrossRef]

37. Joly, P.; Duda, G.N.; Schöne, M.; Welzel, P.B.; Freudenberg, U.; Werner, C.; Petersen, A. Geometry-driven cell organization determines tissue growths in scaffold pores: Consequences for fibronectin organization. *PLoS ONE* **2013**, *8*, e73545. [CrossRef] [PubMed]

38. Drexler, J.W.; Powell, H.M. Regulation of electrospun scaffold stiffness via coaxial core diameter. *Acta Biomater.* **2011**, *7*, 1133–1139. [CrossRef] [PubMed]

39. Vatankhah, E.; Prabhakaran, M.P.; Semnani, D.; Razavi, S.; Zamani, M.; Ramakrishna, S. Phenotypic modulation of smooth muscle cells by chemical and mechanical cues of electrospun tecophilic/gelatin nanofibers. *ACS Appl. Mater. Interfaces* **2014**, *6*, 4089–4101. [CrossRef] [PubMed]

40. Sundararaghavan, H.G.; Monteiro, G.A.; Lapin, N.A.; Chabal, Y.J.; Miksan, J.R.; Shreiber, D.I. Genipin-induced changes in collagen gels: Correlation of mechanical properties to fluorescence. *J. Biomed. Mater. Res. A* **2008**, *87*, 308–320. [CrossRef] [PubMed]

41. Elder, B.D.; Mohan, A.; Athanasiou, K.A. Beneficial effects of exogenous crosslinking agents on self-assembled tissue engineered cartilage construct biomechanical properties. *J. Mech. Med. Biol.* **2011**, *11*, 433–443. [CrossRef] [PubMed]

42. Hughes, C.S.; Postovit, L.M.; Lajoie, G.A. Matrigel: A complex protein mixture required for optimal growth of cell culture. *Proteomics* **2010**, *10*, 1886–1890. [CrossRef] [PubMed]

43. Lancaster, M.A.; Knoblich, J.A. Generation of cerebral organoids from human pluripotent stem cells. *Nat. Protoc.* **2014**, *9*, 2329–2340. [CrossRef] [PubMed]

44. Fraley, S.I.; Feng, Y.; Krishnamurthy, R.; Kim, D.H.; Celedon, A.; Longmore, G.D., Wirtz, D. A distinctive role for focal adhesion proteins in three-dimensional cell motility. *Nat. Cell Biol.* **2010**, *12*, 598–604. [CrossRef] [PubMed]

45. Harunaga, J.S.; Yamada, K.M. Cell-matrix adhesions in 3D. *Matrix Biol.* **2011**, *30*, 363–368. [CrossRef] [PubMed]

46. Kubow, K.E.; Conrad, S.K.; Horwitz, A.R. Matrix microarchitecture and myosin II determine adhesion in 3D matrices. *Curr. Biol.* **2013**, *23*, 1607–1619. [CrossRef] [PubMed]

47. Maheshwari, G.; Brown, G.; Lauffenburger, D.A.; Wells, A.; Griffith, L.G. Cell Adhesion and Motility Depend on Nanoscale RGD Clustering. *J. Cell Sci.* **2000**, *113*, 1677–1686. [PubMed]

48. Nelson, C.M. Geometric control of tissue morphogenesis. *Biochim. Biophys. Acta* **2009**, *1793*, 903–910. [CrossRef] [PubMed]

49. Haeger, A.; Wolf, K.; Zegers, M.M.; Friedl, P. Collective cell migration: Guidance principles and hierarchies. *Trends Cell Biol.* **2015**, *25*, 556–566. [CrossRef] [PubMed]

50. Wu, J.; Mao, Z.; Tan, H.; Han, L.; Ren, T.; Gao, C. Gradient biomaterials and their influences on cell migration. *Interface Focus.* **2012**, *2*, 337–355. [CrossRef] [PubMed]

51. Lee, E.J.; Chan, E.W.L.; Luoa, W.; Yousaf, M.N. Ligand slope, density and affinity direct cell polarity and migration on molecular gradient surfaces. *RSC Adv.* **2014**, *4*, 31581. [CrossRef]

52. Smith, J.T.; Elkin, J.T.; Reichert, W.M. Directed cell migration on fibronectin gradients: Effect of gradient slope. *Exp. Cell Res.* **2006**, *312*, 2424–2432. [CrossRef] [PubMed]

53. Guarnieri, D.; De Capua, A.; Ventre, M.; Borzacchiello, A.; Pedone, C.; Marasco, D.; Ruvo, M.; Netti, P.A. Covalently immobilized RGD gradient on PEG hydrogel scaffold influences cell migration parameters. *Acta Biomater.* **2010**, *6*, 2532–2539. [CrossRef] [PubMed]

54. Lagunas, A.; Martínez, E.; Samitier, J. Surface-Bound Molecular Gradients for the High-Throughput Screening of Cell Responses. *Front. Bioeng. Biotechnol.* **2015**, *3*, 132. [CrossRef] [PubMed]

55. Qin, D.; Xia, Y.; Whitesides, G.M. Soft lithography for micro- and nanoscale patterning. *Nat. Protoc.* **2010**, *5*, 491–502. [CrossRef] [PubMed]

56. Lipomi, D.J.; Martinez, R.V.; Cademartiri, L.; Whitesides, G.M. Soft Lithographic Approaches to Nanofabrication. In *Polymer Science: A Comprehensive Reference*; Matyjaszewski, K., Moller, M., Eds.; Elsevier BV: Amsterdam, The Netherlands, 2012; Volume 7, pp. 211–231.

57. Ventre, M.; Natale, C.F.; Rianna, C.; Netti, P.A. Topographic cell instructive patterns to control cell adhesion, polarization and migration. *J. R. Soc. Interface* **2014**, *11*, 20140687. [CrossRef] [PubMed]

58. Loesberg, W.A.; Te Riet, J.; van Delft, F.C.; Schön, P.; Figdor, C.G.; Speller, S.; van Loon, J.J.; Walboomers, X.F.; Jansen, J.A. The threshold at which substrate nanogroove dimensions may influence fibroblast alignment and adhesion. *Biomaterials* **2007**, *28*, 3944–3951. [CrossRef] [PubMed]

59. Natale, C.F.; Ventre, M.; Netti, P.A. Tuning the material-cytoskeleton crosstalk via nanoconfinement of focal adhesions. *Biomaterials* **2014**, *35*, 2743–2751. [CrossRef] [PubMed]

60. Gerecht, S.; Bettinger, C.J.; Zhang, Z.; Borenstein, J.T.; Vunjak-Novakovic, G.; Langer, R. The Effect of Actin Disrupting Agents on Contact Guidance of Human Embryonic Stem Cells. *Biomaterials* **2007**, *28*, 4068–4077. [CrossRef] [PubMed]

61. Yim, E.K.; Pang, S.W.; Leong, K.W. Synthetic Nanostructures Inducing Differentiation of Human Mesenchymal Stem Cells into Neuronal Lineage. *Exp. Cell Res.* **2007**, *313*, 1820–1829. [CrossRef] [PubMed]

62. Bettinger, C.J.; Langer, R.; Borenstein, J.T. Engineering Substrate Topography at the Micro- and Nanoscale to Control Cell Function. *Angew. Chem. Int. Ed. Engl.* **2009**, *48*, 5406–5415. [CrossRef] [PubMed]

63. Csucs, G.; Michel, R.; Lussi, J.W.; Textor, M.; Danuser, G. Microcontact printing of novel co-polymers in combination with proteins for cell-biological applications. *Biomaterials* **2003**, *24*, 1713–1720. [CrossRef]

64. Eichinger, C.D.; Hsiao, T.W.; Hlady, V. Multiprotein microcontact printing with micrometer resolution. *Langmuir* **2012**, *28*, 2238–2243. [CrossRef] [PubMed]

65. Ye, F.; Jiang, J.; Chang, H.; Xie, L.; Deng, J.; Ma, Z.; Yuan, W. Improved single-cell culture achieved using micromolding in capillaries technology coupled with poly (HEMA). *Biomicrofluidics* **2015**, *9*, 044106. [CrossRef] [PubMed]

66. Marel, A.K.; Rappl, S.; Piera Alberola, A.; Rädler, J.O. Arraying cell cultures using PEG-DMA micromolding in standard culture dishes. *Macromol. Biosci.* **2013**, *13*, 595–602. [CrossRef] [PubMed]

67. Schvartzman, M.; Palma, M.; Sable, J.; Abramson, J.; Hu, X.; Sheetz, M.P.; Wind, S.J. Nanolithographic Control of the Spatial Organization of Cellular Adhesion Receptors at the Single-Molecule Level. *Nano Lett.* **2011**, *11*, 1306–1312. [CrossRef] [PubMed]

68. Glass, R.; Möller, M.; Spatz, J.P. Block Copolymer Micelle Nanolithography. *Nanotechnology* **2003**, *14*, 1153–1160. [CrossRef]

69. Arnold, M.; Cavalcanti-Adam, E.A.; Glass, R.; Blümmel, J.; Eck, W.; Kantlehner, M.; Kessler, H.; Spatz, J.P. Activation of integrin function by nanopatterned adhesive interfaces. *Chemphyschem* **2004**, *5*, 383–388. [CrossRef] [PubMed]

70. Arnold, M.; Hirschfeld-Warneken, V.C.; Lohmüller, T.; Heil, P.; Blümmel, J.; Cavalcanti-Adam, E.A.; López-García, M.; Walther, P.; Kessler, H.; Geiger, B.; *et al.* Induction of cell polarization and migration by a gradient of nanoscale variations in adhesive ligand spacing. *Nano Lett.* **2008**, *8*, 2063–2069. [CrossRef] [PubMed]

71. Altrock, E.; Muth, C.A.; Klein, G.; Spatz, J.P.; Lee-Thedieck, C. The significance of integrin ligand nanopatterning on lipid raft clustering in hematopoietic stem cells. *Biomaterials* **2012**, *33*, 3107–3118. [CrossRef] [PubMed]

72. Nimmo, C.M.; Shoichet, M.S. Regenerative biomaterials that "click": Simple, aqueous-based protocols for hydrogel synthesis, surface immobilization, and 3D patterning. *Bioconjug. Chem.* **2011**, *22*, 2199–2209. [CrossRef] [PubMed]

73. Kharkar, P.M.; Kiick, K.L.; Kloxin, A.M. Designing degradable hydrogels for orthogonal control of cell microenvironments. *Chem. Soc. Rev.* **2013**, *42*, 7335–7972. [CrossRef] [PubMed]

74. Ifkovits, J.L.; Burdick, J.A. Review: Photopolymerizable and degradable biomaterials for tissue engineering applications. *Tissue Eng.* **2007**, *13*, 2369–2385. [CrossRef] [PubMed]

75. Fairbanks, B.D.; Schwartz, M.P.; Bowman, C.N.; Anseth, K.S. Photoinitiated polymerization of PEG-diacrylate with lithium phenyl-2,4,6-trimethylbenzoylphosphinate: Polymerization rate and cytocompatibility. *Biomaterials* **2009**, *30*, 6702–6707. [CrossRef] [PubMed]

76. Lutolf, M.P.; Lauer-Fields, J.L.; Schmoekel, H.G.; Metters, A.T.; Weber, F.E.; Fields, G.B.; Hubbell, J.A. Synthetic matrix metalloproteinase-sensitive hydrogels for the conduction of tissue regeneration: Engineering cell-invasion characteristics. *Proc. Natl. Acad. Sci. USA* **2003**, *100*, 5413–5418. [CrossRef] [PubMed]

77. Lauffenburger, D.A.; Horwitz, A.F. Cell migration: A physically integrated molecular process. *Cell* **1996**, *84*, 359–369. [CrossRef]

78. Sanborn, T.J.; Messersmith, P.B.; Barron, A.E. In situ crosslinking of a biomimetic peptide-PEG hydrogel via thermally triggered activation of factor XIII. *Biomaterials* **2002**, *23*, 2703–2710. [CrossRef]

79. Ehrbar, M.; Rizzi, S.C.; Schoenmakers, R.G.; Miguel, B.S.; Hubbell, J.A.; Weber, F.E.; Lutolf, M.P. Biomolecular hydrogels formed and degraded via site-specific enzymatic reactions. *Biomacromolecules* **2007**, *8*, 3000–3007. [CrossRef] [PubMed]

80. Oliviero, O.; Ventre, M.; Netti, P.A. Functional porous hydrogels to study angiogenesis under the effect of controlled release of vascular endothelial growth factor. *Acta Biomater.* **2012**, *8*, 3294–3301. [CrossRef] [PubMed]

81. Li, Z.; Qu, T.; Ding, C.; Ma, C.; Sun, H.; Li, S.; Liu, X. Injectable gelatin derivative hydrogels with sustained vascular endothelial growth factor release for induced angiogenesis. *Acta Biomater.* **2015**, *13*, 88–100. [CrossRef] [PubMed]

82. Zisch, A.H.; Lutolf, M.P.; Ehrbar, M.; Raeber, G.P.; Rizzi, S.C.; Davies, N.; Schmökel, H.; Bezuidenhout, D.; Djonov, V.; Zilla, P.; *et al.* Cell-demanded release of VEGF from synthetic, biointeractive cell ingrowth matrices for vascularized tissue growth. *FASEB J.* **2003**, *17*, 2260–2262. [CrossRef] [PubMed]

83. Nelson, C.M.; Vanduijn, M.M.; Inman, J.L.; Fletcher, D.A.; Bissell, M.J. Tissue geometry determines sites of mammary branching morphogenesis in organotypic cultures. *Science* **2006**, *314*, 298–300. [CrossRef] [PubMed]

84. Khetan, S.; Burdick, J.A. Patterning hydrogels in three dimensions towards controlling cellular interactions. *Soft Matter* **2011**, *7*, 830–838. [CrossRef]

85. Pereira, R.F.; Bártolo, P.J. 3D Photo-Fabrication for Tisse Engineering and Drug Delivery. *Engineering* **2015**, *1*, 90–112. [CrossRef]

86. Curley, J.L.; Moore, M.J. Facile micropatterning of dual hydrogel systems for 3D models of neurite outgrowth. *J. Biomed. Mater. Res. A.* **2011**, *99*, 532–543. [CrossRef] [PubMed]

87. Itoga, K.; Kobayashi, J.; Yamato, M.; Kikuchi, A.; Okano, T. Maskless liquid-crystal-display projection photolithography for improved design flexibility of cellular micropatterns. *Biomaterials* **2006**, *27*, 3005–3009. [CrossRef] [PubMed]

88. Hanson Shepherd, J.N.; Parker, S.T.; Shepherd, R.F.; Gillette, M.U.; Lewis, J.A.; Nuzzo, R.G. 3D Microperiodic Hydrogel Scaffolds for Robust Neuronal Cultures. *Adv. Funct. Mater.* **2011**, *21*, 47–54. [CrossRef] [PubMed]

89. Liu Tsang, V.; Chen, A.A.; Cho, L.M.; Jadin, K.D.; Sah, R.L.; DeLong, S.; West, J.L.; Bhatia, S.N. Fabrication of 3D hepatic tissues by additive photopatterning of cellular hydrogels. *FASEB J.* **2007**, *21*, 790–801. [CrossRef] [PubMed]

90. Schuster, M.; Turecek, C.; Weigel, G.; Saf, R.; Stampfl, J.; Varga, F.; Liska, R. Gelatin-based photopolymers for bone replacement materials. *J. Polym. Sci. A Polym. Chem.* **2009**, *47*, 7078–7089. [CrossRef]

91. Zorlutuna, P.; Jeong, H.J.; Kong, H.; Bashir, R. Stereolithography-Based Hydrogel Microenvironments to Examine Cellular Interactions. *Adv. Funct. Mater.* **2011**, *21*, 3642–3651. [CrossRef]

92. Ovsianikov, A.; Gruene, M.; Pflaum, M.; Koch, L.; Maiorana, F.; Wilhelmi, M.; Haverich, A.; Chichkov, B. Laser printing of cells into 3D scaffolds. *Biofabrication* **2010**, *2*, 014104. [CrossRef] [PubMed]

93. Ovsianikov, A.; Deiwick, A.; Van Vlierberghe, S.; Dubruel, P.; Möller, L.; Dräger, G.; Chichkov, B. Laser fabrication of three-dimensional CAD scaffolds from photosensitive gelatin for applications in tissue engineering. *Biomacromolecules* **2011**, *12*, 851–858. [CrossRef] [PubMed]

94. Chan, V.; Zorlutuna, P.; Jeong, J.H.; Kong, H.; Bashir, R. Three-dimensional photopatterning of hydrogels using stereolithography for long-term cell encapsulation. *Lab Chip* **2010**, *10*, 2062–2070. [CrossRef] [PubMed]

95. Tirella, A.; Orsini, A.; Vozzi, G.; Ahluwalia, A. A phase diagram for microfabrication of geometrically controlled hydrogel scaffolds. *Biofabrication* **2009**, *1*, 045002. [CrossRef] [PubMed]

96. Xu, M.; Wang, X.; Yan, Y.; Yao, R.; Ge, Y. An cell-assembly derived physiological 3D model of the metabolic syndrome, based on adipose-derived stromal cells and a gelatin/alginate/fibrinogen matrix. *Biomaterials* **2010**, *31*, 3868–3877. [CrossRef] [PubMed]

97. Miller, J.S.; Stevens, K.R.; Yang, M.T.; Baker, B.M.; Nguyen, D.H.; Cohen, D.M.; Toro, E.; Chen, A.A.; Galie, P.A.; Yu, X.; *et al.* Rapid casting of patterned vascular networks for perfusable engineered three-dimensional tissues. *Nat. Mater.* **2012**, *11*, 768–774. [CrossRef] [PubMed]

98. Hribar, K.C.; Meggs, K.; Liu, J.; Zhu, W.; Qu, X.; Chen, S. Three-dimensional direct cell patterning in collagen hydrogels with near-infrared femtosecond laser. *Sci. Rep.* **2015**, *5*, 17203. [CrossRef] [PubMed]

99. Kloxin, A.M.; Kasko, A.M.; Salinas, C.N.; Anseth, K.S. Photodegradable hydrogels for dynamic tuning of physical and chemical properties. *Science* **2009**, *324*, 59–63. [CrossRef] [PubMed]

100. Kloxin, A.M.; Tibbitt, M.W.; Kasko, A.M.; Fairbairn, J.A.; Anseth, K.S. Tunable hydrogels for external manipulation of cellular microenvironments through controlled photodegradation. *Adv. Mater.* **2010**, *22*, 61–66. [CrossRef] [PubMed]

101. Lee, S.H.; Moon, J.J.; West, J.L. Three-dimensional micropatterning of bioactive hydrogels via two-photon laser scanning photolithography for guided 3D cell migration. *Biomaterials* **2008**, *29*, 2962–2968. [CrossRef] [PubMed]

102. Lee, W.; Debasitis, J.C.; Lee, V.K.; Lee, J.H.; Fischer, K.; Edminster, K.; Park, J.K.; Yoo, S.S. Multi-layered culture of human skin fibroblasts and keratinocytes through three-dimensional freeform fabrication. *Biomaterials* **2009**, *30*, 1587–1595. [CrossRef] [PubMed]

103. Billiet, T.; Vandenhaute, M.; Schelfhout, J.; Van Vlierberghe, S.; Dubruel, P. A review of trends and limitations in hydrogel-rapid prototyping for tissue engineering. *Biomaterials* **2012**, *33*, 6020–6041. [CrossRef] [PubMed]

104. Lam, C.X.F.; Mo, X.M.; Teoh, S.H.; Hutmacher, D.W. Scaffold development using 3D printing with a starch-based polymer. *Mater. Sci. Eng. C Mater. Biol. Appl.* **2002**, *20*, 49–56. [CrossRef]

105. Xu, T.; Gregory, C.A.; Molnar, P.; Cui, X.; Jalota, S.; Bhaduri, S.B.; Boland, T. Viability and electrophysiology of neural cell structures generated by the inkjet printing method. *Biomaterials* **2006**, *27*, 3580–3588. [CrossRef] [PubMed]

106. Norotte, C.; Marga, F.S.; Niklason, L.E.; Forgacs, G. Scaffold-free vascular tissue engineering using bioprinting. *Biomaterials* **2009**, *30*, 5910–5917. [CrossRef] [PubMed]

107. Jakab, K.; Norotte, C.; Damon, B.; Marga, F.; Neagu, A.; Besch-Williford, C.L.; Kachurin, A.; Church, K.H.; Park, H.; Mironov, V.; *et al.* Tissue engineering by self-assembly of cells printed into topologically defined structures. *Tissue Eng. Part. A.* **2008**, *14*, 413–421. [CrossRef] [PubMed]

108. Owens, C.M.; Marga, F.; Forgacs, G.; Heesch, C.M. Biofabrication and testing of a fully cellular nerve graft. *Biofabrication* **2013**, *5*, 045007. [CrossRef] [PubMed]

109. Eyckmans, J.; Boudou, T.; Yu, X.; Chen, C.S. A hitchhiker's guide to mechanobiology. *Dev. Cell.* **2011**, *21*, 35–47. [CrossRef] [PubMed]

110. Bershadsky, A.D.; Ballestrem, C.; Carramusa, L.; Zilberman, Y.; Gilquin, B.; Khochbin, S.; Alexandrova, A.Y.; Verkhovsky, A.B.; Shemesh, T.; Kozlov, M.M. Assembly and mechanosensory function of focal adhesions: Experiments and models. *Eur. J. Cell Biol.* **2006**, *85*, 165–173. [CrossRef] [PubMed]

111. Houseman, B.T.; Mrksich, M. The microenvironment of immobilized Arg-Gly-Asp peptides is an important determinant of cell adhesion. *Biomaterials* **2001**, *22*, 943–955. [CrossRef]

112. Kuhlman, W.; Taniguchi, I.; Griffith, L.G.; Mayes, A.M. Interplay between PEO tether length and ligand spacing governs cell spreading on RGD-modified PMMA-g-PEO comb copolymers. *Biomacromolecules* **2007**, *8*, 3206–3213. [CrossRef] [PubMed]

113. Choi, C.K.; Xu, Y.J.; Wang, B.; Zhu, M.; Zhang, L.; Bian, L. Substrate Coupling Strength of Integrin-Binding Ligands Modulates Adhesion, Spreading, and Differentiation of Human Mesenchymal Stem Cells. *Nano Lett.* **2015**, *15*, 6592–6600. [CrossRef] [PubMed]

114. Trappmann, B.; Gautrot, J.E.; Connelly, J.T.; Strange, D.G.; Li, Y.; Oyen, M.L.; Cohen Stuart, M.A.; Boehm, H.; Li, B.; Vogel, V.; *et al.* Extracellular-matrix tethering regulates stem-cell fate. *Nat. Mater.* **2012**, *11*, 642–649. [CrossRef] [PubMed]

115. Wen, J.H.; Vincent, L.G.; Fuhrmann, A.; Choi, Y.S.; Hribar, K.C.; Taylor-Weiner, H.; Chen, S.; Engler, A.J. Interplay of matrix stiffness and protein tethering in stem cell differentiation. *Nat. Mater.* **2014**, *13*, 979–987. [CrossRef] [PubMed]

116. Fu, J.; Wang, Y.K.; Yang, M.T.; Desai, R.A.; Yu, X.; Liu, Z.; Chen, C.S. Mechanical regulation of cell function with geometrically modulated elastomeric substrates. *Nat. Methods.* **2010**, *7*, 733–736. [CrossRef] [PubMed]

117. Engler, A.J.; Sen, S.; Sweeney, H.L.; Discher, D.E. Matrix elasticity directs stem cell lineage specification. *Cell* **2006**, *126*, 677–589. [CrossRef] [PubMed]

118. Kilian, K.A.; Bugarija, B.; Lahn, B.T.; Mrksich, M. Geometric Cues for Directing the Differentiation of Mesenchymal Stem Cells. *Proc. Natl. Acad. Sci. U. S. A.* **2010**, *107*, 4872–4877. [CrossRef] [PubMed]

119. Peng, R.; Yao, X.; Ding, J. Effect of Cell Anisotropy on Differentiation of Stem Cells on Micropatterned Surfaces through the Controlled Single Cell Adhesion. *Biomaterials* **2011**, *32*, 8048–8057. [CrossRef] [PubMed]

120. Yao, X.; Peng, R.; Ding, J. Effects of Aspect Ratios of Stem Cells on Lineage Commitments with and without Induction Media. *Biomaterials* **2013**, *34*, 930–939. [CrossRef] [PubMed]

121. Wang, X.; Yan, C.; Ye, K.; He, Y.; Li, Z.; Ding, J. Effect of RGD nanospacing on differentiation of stem cells. Effect of RGD nanospacing on differentiation of stem cells. *Biomaterials* **2013**, *34*, 2865–2874. [CrossRef] [PubMed]

122. Young, J.L.; Engler, A.J. Hydrogels with time-dependent material properties enhance cardiomyocyte differentiation *in vitro*. *Biomaterials* **2011**, *32*, 1002–1009. [CrossRef] [PubMed]

123. Cameron, A.R.; Frith, J.E.; Cooper-White, J.J. The influence of substrate creep on mesenchymal stem cell behaviour and phenotype. *Biomaterials* **2011**, *32*, 5979–5993. [CrossRef] [PubMed]

124. Cameron, A.R.; Frith, J.E.; Gomez, G.A.; Yap, A.S.; Cooper-White, J.J. The effect of time-dependent deformation of viscoelastic hydrogels on myogenic induction and Rac1 activity in mesenchymal stem cells. *Biomaterials* **2014**, *35*, 1857–1868. [CrossRef] [PubMed]

125. Chaudhuri, O.; Gu, L.; Darnell, M.; Klumpers, D.; Bencherif, S.A.; Weaver, J.C.; Huebsch, N.; Mooney, D.J. Substrate stress relaxation regulates cell spreading. *Nat. Commun.* **2015**, *6*, 6364. [CrossRef] [PubMed]

126. Eppell, S.J.; Smith, B.N.; Kahn, H.; Ballarini, R. Nano measurements with micro-devices: Mechanical properties of hydrated collagen fibrils. *J. R. Soc. Interface* **2006**, *3*, 117–121. [CrossRef] [PubMed]

127. Paszek, M.J.; Zahir, N.; Johnson, K.R.; Lakins, J.N.; Rozenberg, G.I.; Gefen, A.; Reinhart-King, C.A.; Margulies, S.S.; Dembo, M.; Boettiger, D.; *et al.* Tensional homeostasis and the malignant phenotype. *Cancer Cell* **2005**, *8*, 241–254. [CrossRef] [PubMed]

128. Ahearne, M.; Liu, K.K.; El Haj, A.J.; Then, K.Y.; Rauz, S.; Yang, Y. Online monitoring of the mechanical behavior of collagen hydrogels: Influence of corneal fibroblasts on elastic modulus. *Tissue Eng. C* **2010**, *16*, 319–327. [CrossRef] [PubMed]

129. Tranquillo, R.T. Self-organization of tissue-equivalents: The nature and role of contact guidance. *Biochem. Soc. Symp.* **1999**, *65*, 27–42. [PubMed]

130. Huebsch, N.; Arany, P.R.; Mao, A.S.; Shvartsman, D.; Ali, O.A.; Bencherif, S.A.; Rivera-Feliciano, J.; Mooney, D.J. Harnessing traction-mediated manipulation of the cell/matrix interface to control stem-cell fate. *Nat. Mater.* **2010**, *9*, 518–526. [CrossRef] [PubMed]

131. Khetan, S.; Burdick, J.A. Patterning network structure to spatially control cellular remodeling and stem cell fate within 3-dimensional hydrogels. *Biomaterials* **2010**, *31*, 8228–8234. [CrossRef] [PubMed]

132. Guvendiren, M.; Burdick, J.A. Stiffening hydrogels to probe short- and long-term cellular responses to dynamic mechanics. *Nat. Commun.* **2012**, *3*, 792. [CrossRef] [PubMed]

133. Khetan, S.; Guvendiren, M.; Legant, W.R.; Cohen, D.M.; Chen, C.S.; Burdick, J.A. Degradation-mediated cellular traction directs stem cell fate in covalently crosslinked three-dimensional hydrogels. *Nat. Mater.* **2013**, *12*, 458–465. [CrossRef] [PubMed]

134. Zhang, S.; Holmes, T.; Lockshin, C.; Rich, A. Spontaneous assembly of a self-complementary oligopeptide to form a stable macroscopic membrane. *Proc. Natl. Acad. Sci. USA* **1993**, *90*, 3334–3338. [CrossRef] [PubMed]

135. Ulijn, R.V.; Smith, A.M. Designing peptide based nanomaterials. *Chem. Soc. Rev.* **2008**, *37*, 664–675. [CrossRef] [PubMed]

136. Papapostolou, D.; Smith, A.M.; Atkins, E.D.; Oliver, S.J.; Ryadnov, M.G.; Serpell, L.C.; Woolfson, D.N. Engineering nanoscale order into a designed protein fiber. *Proc. Natl. Acad. Sci. USA* **2007**, *104*, 10853–10858. [CrossRef] [PubMed]

137. Pochan, D.J.; Schneider, J.P.; Kretsinger, J.; Ozbas, B.; Rajagopal, K.; Haines, L. Thermally reversible hydrogels via intramolecular folding and consequent self-assembly of a de novo designed peptide. *J. Am. Chem. Soc.* **2003**, *125*, 11802–11803. [CrossRef] [PubMed]

138. Ramachandran, S.; Tseng, Y.; Yu, Y.B. Repeated rapid shear-responsiveness of peptide hydrogels with tunable shear modulus. *Biomacromolecules* **2005**, *6*, 1316–1321. [CrossRef] [PubMed]

139. Holmes, T.C.; de Lacalle, S.; Su, X.; Liu, G.; Rich, A.; Zhang, S. Extensive neurite outgrowth and active synapse formation on self-assembling peptide scaffolds. *Proc. Natl. Acad. Sci. USA* **2000**, *97*, 6728–6733. [CrossRef] [PubMed]

140. Kumada, Y.; Zhang, S. Significant type I and type III collagen production from human periodontal ligament fibroblasts in 3D peptide scaffolds without extra growth factors. *PLoS ONE* **2010**, *5*, e10305. [CrossRef] [PubMed]

141. Kumada, Y.; Hammond, N.A.; Zhang, S.G. Functionalized scaffolds of shorter selfassembling peptides containing MMP-2 cleavable motif promote fibroblast proliferation and significantly accelerate 3-D cell migration independent of scaffold stiffness. *Soft Matter* **2010**, *6*, 5073–5079. [CrossRef]

142. Galler, K.M.; Aulisa, L.; Regan, K.R.; D'Souza, R.N.; Hartgerink, J.D. Self-assembling multidomain peptide hydrogels: Designed susceptibility to enzymatic cleavage allows enhanced cell migration and spreading. *J. Am. Chem. Soc.* **2010**, *132*, 3217–3223. [CrossRef] [PubMed]

143. Cui, H.; Webber, M.J.; Stupp, S.I. Self-assembly of peptide amphiphiles: From molecules to nanostructures to biomaterials. *Biopolymers* **2010**, *94*, 1–18. [CrossRef] [PubMed]

144. Pashuck, E.T.; Cui, H.; Stupp, S.I. Tuning supramolecular rigidity of peptide fibers through molecular structure. *J. Am. Chem. Soc.* **2010**, *5*, 132. [CrossRef] [PubMed]

145. Boekhoven, J.; Stupp, S.I. 25th anniversary article: Supramolecular materials for regenerative medicine. *Adv. Mater.* **2014**, *26*, 1642–1659. [CrossRef] [PubMed]

146. Storrie, H.; Guler, M.O.; Abu-Amara, S.N.; Volberg, T.; Rao, M.; Geiger, B.; Stupp, S.I. Supramolecular crafting of cell adhesion. *Biomaterials* **2007**, *28*, 4608–4618. [CrossRef] [PubMed]

147. Silva, G.A.; Czeisler, C.; Niece, K.L.; Beniash, E.; Harrington, D.A.; Kessler, J.A.; Stupp, S.I. Selective differentiation of neural progenitor cells by high-epitope density nanofibers. *Science* **2004**, *303*, 1352–1355. [CrossRef] [PubMed]

148. Yang, Z.; Liang, G.; Wang, L.; Xu, B. Using a kinase/phosphatase switch to regulate a supramolecular hydrogel and forming the supramolecular hydrogel *in vivo*. *J. Am. Chem. Soc.* **2006**, *128*, 3038–3043. [CrossRef] [PubMed]

149. Lee, H.K.; Soukasene, S.; Jiang, H.; Zhang, S.; Feng, W.; Stupp, S.I. Light-induced self-assembly of nanofibers inside liposomes. *Soft Matter* **2008**, *4*, 962–964. [CrossRef] [PubMed]

150. Muraoka, T.; Koh, C.Y.; Cui, H.; Stupp, S.I. Light-triggered bioactivity in three dimensions. *Angew. Chem. Int. Ed. Engl.* **2009**, *48*, 5946–5949. [CrossRef] [PubMed]

151. Gjorevski, N.; Ranga, A.; Lutolf, M.P. Bioengineering approaches to guide stem cell-based organogenesis. *Development* **2014**, *141*, 1794–1804. [CrossRef] [PubMed]

152. Iannone, M.; Ventre, M.; Formisano, L.; Casalino, L.; Patriarca, E.J.; Netti, P.A. Nanoengineered surfaces for focal adhesion guidance trigger mesenchymal stem cell self-organization and tenogenesis. *Nano Lett.* **2015**, *15*, 1517–1525. [CrossRef] [PubMed]

153. Gobaa, S.; Hoehnel, S.; Roccio, M.; Negro, A.; Kobel, S.; Lutolf, M.P. Artificial niche microarrays for probing single stem cell fate in high throughput. *Nat. Methods.* **2011**, *8*, 949–955. [CrossRef] [PubMed]

154. Roccio, M.; Gobaa, S.; Lutolf, M.P. High-throughput clonal analysis of neural stem cells in microarrayed artificial niches. *Integr. Biol. (Camb).* **2012**, *4*, 391–400. [CrossRef] [PubMed]

155. Floren, M.; Tan, W. Three-dimensional, soft neotissue arrays as high throughput platforms for the interrogation of engineered tissue environments. *Biomaterials* **2015**, *59*, 39–52. [CrossRef] [PubMed]

156. Vats, K.; Benoit, D.S. Dynamic manipulation of hydrogels to control cell behavior: A review. *Tissue Eng B Rev.* **2013**, *19*, 455–469. [CrossRef] [PubMed]

157. Huebsch, N.; Lippens, E.; Lee, K.; Mehta, M.; Koshy, S.T.; Darnell, M.C.; Desai, R.M.; Madl, C.M.; Xu, M.; Zhao, X.; *et al.* Matrix elasticity of void-forming hydrogels controls transplanted-stem-cell-mediated bone formation. *Nat. Mater.* **2015**, *14*, 1269–1277. [CrossRef] [PubMed]

*Review*

# Combinatorial Method/High Throughput Strategies for Hydrogel Optimization in Tissue Engineering Applications

**Laura A. Smith Callahan [1,2]**

[1] Vivian L. Smith Department of Neurosurgery & Center for Stem Cells and Regenerative Medicine McGovern Medical School at the University of Texas Health Science Center at Houston, Houston, TX 77030, USA; laura.a.smithcallahan@uth.tmc.edu; Tel.: +1-713-500-3431

[2] Department of Nanomedicine and Biomedical Engineering, McGovern Medical School at the University of Texas Health Science Center at Houston, Houston, TX 77030, USA

Academic Editor: Esmaiel Jabbari
Received: 25 March 2016; Accepted: 3 June 2016; Published: 8 June 2016

**Abstract:** Combinatorial method/high throughput strategies, which have long been used in the pharmaceutical industry, have recently been applied to hydrogel optimization for tissue engineering applications. Although many combinatorial methods have been developed, few are suitable for use in tissue engineering hydrogel optimization. Currently, only three approaches (design of experiment, arrays and continuous gradients) have been utilized. This review highlights recent work with each approach. The benefits and disadvantages of design of experiment, array and continuous gradient approaches depending on study objectives and the general advantages of using combinatorial methods for hydrogel optimization over traditional optimization strategies will be discussed. Fabrication considerations for combinatorial method/high throughput samples will additionally be addressed to provide an assessment of the current state of the field, and potential future contributions to expedited material optimization and design.

**Keywords:** tissue engineering; hydrogel; high throughput; design of experiment; array and continuous gradient

## 1. Introduction

The extracellular matrix (ECM) was once thought to be inert [1], but has been found to significantly influence cellular behavior [2,3]. The native ECM of most tissues in the body is a highly hydrated, viscoelastic network of proteoglycans, glycoaminoglycans, and proteins, which provide mechanical, chemical and physical cues to guide cell behavior and tissue homeostasis [1]. A similar progression from being viewed as inert to possessing the potential to guide cellular behavior through compositional blending, tailoring material properties and inclusion of bioactive signaling has occurred with biomaterials [4–8]. Our growing understanding of ECM function in directing cellular behavior has driven this change and spurred a movement to emulate key aspects of the ECM with biomaterials. The complexity of our emulation strategies have advanced with both our biological understanding and technical capabilities. Determining the minimal number of factors needed to achieve the desired cellular behavior outcome remains the major design objective for matrices. Due to their high aqueous content and tailorable material properties that can mimic native ECM properties in many tissues, hydrogels have become widely used as the base matrix in these ECM emulation strategies for tissue engineering platforms [9].

The versatility of hydrogels leads to a number of parameters that can be altered to meet the design criteria for a given tissue engineering application. For instance, hydrogels can be composed of synthetic polymers, natural polymers, or a hybrid of the two [10–12]. The gelation mechanism can be varied from photopolymerization to click chemistry, to Michael addition, to physical entanglement, *etc.* [10,11,13,14]. Bioactive signaling molecules can be tethered or released from the hydrogel at various concentrations and rates [12,15–17]. Even the topography, porosity and crosslink density can be manipulated [18–20]. The sheer number of design options for hydrogel construction has traditionally led to an ad hoc trial and error approach to their testing and optimization for clinical use. This has led to a limited understanding of key hydrogel design parameters that affect cellular behavior, slowed the development process and reduced the number of hydrogel based treatments reaching clinical application.

Combinatorial method/high throughput strategies, which have long been used in pharmaceutical development to screen multiple molecules in parallel [21], offer the potential to expedite hydrogel development for clinical tissue engineering applications and our understanding of cell-biomaterial interfaces. These strategies allow for efficient exploration of a large compositional space [22]. They are particularly useful as the ability to theoretically predict cellular response to hydrogels is currently limited, meaning brute force experimentation is necessary to develop enough base information for development of robust predictive models [22]. If coupled with sample miniaturization and automation, these strategies can reduce the overall cost of the research in terms of reagents and manpower. The number of cells needed to obtain the information can also be reduced, which is particularly helpful with rare or difficult to culture cell types, compared to traditional approaches.

Although many combinatorial method/high throughput strategies have been developed [23–25], few are suitable for hydrogel optimization in tissue engineering applications as both two- and three-dimensional cell culture are necessary and multiple types of design parameters ranging from composition to functionalization with bioactive signaling molecules must be examined. Currently, the combinatorial/high throughput strategies that have been applied to hydrogel optimization are design of experiment (DOE), arrays and continuous gradients. In this review, the basics of each of the applied methods will be covered along with the advantages and drawbacks. Since these strategies could ultimately be used in combination for hydrogel optimization, general sample design considerations will be covered as well.

## 2. Design of Experiment

DOE is a statistical software approach used to identify the best parameters for a desired outcome. In this method, each input is entered into a predictive model in order to determine the combination of factors necessary to achieve the desired outcome. Normally multiple rounds of experiments are necessary to optimize the material and achieve the cellular response desired as the data from the current round of optimizations is added to the predictive model to refine the predictions for the next round of testing. The number of rounds is dependent on the number of inputs, outcomes and the design approach, either full or fractional factorial [26]. Full factorial design covers all the possible combinations of the inputs. This requires lots of resources (manpower and reagents), which makes it impractical for many studies. Fractional design runs the minimum number of experiments to identify the desired outcome, so only a portion of the possible combinations are run. In this case, the needed resources are reduced, but the potential to misidentify or completely miss effects are possible. Even with reduced resource requirements, the number of samples necessary for the optimization can still be large in order to identify the main effects and crosstalk interactions. One study optimizing 35 test conditions, ranging from cell density and ratio of different cell types to hydrogel composition and thickness, for human umbilical vein endothelial cell (HUVEC) culture required 200 samples to identify a paracrine effect between vascular network formation and osteogenically differentiated human mesenchymal stem cells (hMSC) and create a three dimensional hydrogel model for further biological study of this effect [27].

Fractional DOE design has been used to optimize Arg-Gly-Asp (RGD), Tyr-Ile-Gly-Ser-Arg (YIGSR) and Ile-Lys-Val-Ala-Val (IKVAV) peptide concentration simultaneously in hydrogels instead of more traditional approach of optimizing one peptide at a time. One study optimizing the RGD (8 mM), YIGSR (0 mM) and IKVAV (3 mM) peptide concentrations for HUVEC culture in nanofibrous self-assembly peptide scaffolds identified an antagonistic relation between RGD and YIGSR, contrary to previous reports that had not utilized a high throughput strategy [28]. Another study optimized the RGD (100 µM), YIGSR (48 µM) and IKVAV (300 µM) peptide concentrations to promote human induced pluripotent stem cells survival during neural differentiation, increasing the number of neural progenitors available for further study [29]. Both studies found simultaneous optimization yielded different optimal concentrations for each peptide with improved biological response compared to individual optimization followed by combination of those peptide concentrations into a single sample. Both studies demonstrate the power of the DOE approach to efficiently optimize multiple design parameters. In addition, its potential to reveal synergistic or detrimental effects due to parameter interactions that may not be uncovered with traditional optimization strategies. However, this efficiency comes from access to the software, which is expensive, and good predictive models. Typically, good predictive models are made using data from experiments examining the cellular response to each test parameter individually. For many design parameters and cell types, the information needed to populate the predictive model is not known and must be obtained, adding to the number of necessary experiments. As the base of knowledge regarding cellular response grows the quality of the initial predictive model will become less of an issue.

## 3. Arrays

Arrays use a number of discrete, often miniaturized, samples to optimize material properties for a desired cellular response. The strategy is compatible with a number of formats. Hydrogel arrays have been printed with soft lithography [30], direct contact printing [22] and inkjet printing [31]; injection molded in microfluidic channels [32]; held in free floating molds [33,34] and attached to glass cover slides [35]. This flexibility allows for their automated fabrication with liquid handling systems. A comparison of cell viability between automated and hand pipetted array systems has found higher cellular viability when automated fabrication was used with multiple cell types [33]. Use of automated liquid handing systems allows for larger arrays with more test conditions than could typically be fabricated by hand. One study, which utilized an automated liquid handling system to fabricate the array, examined 400 test conditions in gelatin hydrogels to study protein effects on hMSC osteogenic differentiation via mineralization of the gelatin matrix [36]. Another study assessing the effects of five different signaling types on the maintenance of pluripotency in mouse embryonic stem cells examined over 1000 test conditions [37]. However, automated liquid handling systems are expensive, which inhibits many from using them to build arrays and significantly lowers the number of test conditions examined in many studies. For comparison, one large array study fabricated by hand pipetting examined 19 test conditions with seven cell types [38]. Reductions in the number of test conditions decreases the chance that optimal conditions will be identified and that secondary relationships or interactions between test parameters will be detected. This mitigates some of the advantages of utilizing a combinatorial/high throughput approach. However, the development of graphical bar codes on polymer and ECM components offers a way to increase tested conditions in hand pipetted arrays [39], making increases in test condition numbers in manually fabricated systems practical. Eventually, the development of less expensive technologies for array fabrication, such hydrophobically created microgels, could additionally ease the burden of hand fabrication or bring the cost of automated array fabrication within the reach of more researchers [40].

Although not ideal for optimization, arrays are well suited for initial discovery of potentially advantageous hydrogel conditions. Inkjet printing coupled with a reduction-oxidation reaction allows for the addition of multiple materials to create complex formulations [31]. Using this approach, a study focused on polymer discovery examined 2280 different formulations to identify a thermally responsive polymer which would release cells upon cooling to room temperature from 37 °C [41]. Use of automated liquid handling for contact printing followed by photopolymerization is another strategy used for this type of array fabrication [22,42]. This approach was used to identify monomers capable of promoting human embryonic stem cell (hESC) differentiation on poly(hydroxyethyl methacrylate) hydrogels from over 1700 test formulations [22].

Alterations in material properties due to changes in composition have been studied using arrays. These studies have ranged from characterizations of gelation kinetics [43] to fibronectin absorption [42]. A study of the effects of polymer chemistry on hESC attachment provided enough data to create a model capable of predicting hESC attachment to novel polymers [44]. The strategy has even been used to identify the optimal hydrogel to release Lipolexe-based transfection agents for efficient transfection of cells cultured on the hydrogel surface [45]. Moving beyond standard material characterization, the flexibility of array construction has allowed for arrays with spatial patterns [46], printed on surfaces with nanofibrous architecture [47] and multiplexed test parameters due to independent patterning methodologies [48] to be fabricated and characterized. This increased fabrication complexity allows for the greater emulation of the native ECM at a structural level and the study of multiple physical stimuli on material behavior and cellular response at the same time.

Cellular response to a number of hydrogel design parameters have been examined with the array format. These studies have focused on identifying optimal hydrogel formulation [22,42,45,49–51], mechanical properties [35,52], hydrogel degradation [33,37,53] and bioactive signaling molecule concentration [33,35,37,52,54] to illicit a desired cellular response. The range of desired cell behaviors observed and used as the selection criteria for the optimal hydrogel has ranged from attachment [41,42,44,51,55–57], cellular morphology [32] and viability [35,48,53,54,58] to migration [52] and lineage choice [30,54]. As the study of cell-biomaterial interface has advanced, so has the selection of cellular behavior utilized as the material selection criteria. This has led to more advanced studies examining the effect of protein concentration on non-adherent neurosphere proliferation, quiescence and death [54], and the ability of dendritic cells to undergo phagocytosis while adhering to hydrogels to be conducted in array format [57].

Like DOE, these studies can optimize more than one parameter at a time. Simultaneous optimization of three parameters at the same time have been reported [37]. One recent study of hMSC adhesion found that changes in Young's modulus altered cellular spreading and focal adhesion formation in response to RGD concentration, indicating an interconnection between the two signals in the cell [35]. However, a similar study of Young's modulus and RGD concentration found the materials mechanical properties to be the major factor affecting fibrosarcoma cellular morphology and migration [52]. The conflicting results between cell types demonstrates the need to run these systematic studies for every cell type of interest as results from one cell type cannot easily be extrapolated to predict the response of another cell type. To further highlight the complexity of parameter interaction that combinatorial methods/high throughput studies can detect, a recent study by Ranga and co-workers examined the effects of Young's Modulus, matrix degradability, tethered and released bioactive signaling molecules, and cellular density on the maintenance of pluripotency in mouse embryonic stem cells using bioinformatic analysis tools [37]. Although their work illustrated to the predominate role of leukemia inhibitory factor in this process, it identified synergistic and detrimental effects of the other test parameters on this process, which had not previously been identified.

## 4. Gradient Samples

Gradient samples can be fabricated with simple inexpensive systems comprised of pumps and molds [59]. Although elimination of the pump is possible through use of passive methods such as surface tension to drive flow through the mold [60]. Newer methods of gradient formation are less reliant on pumps for their formation as they use the mold [61], thermal cycles [61,62], or ultraviolet light exposure [63,64] to generate the gradient instead of flow. Gradient samples consist of a gradual compositional change between two or more parameters (Figure 1). However, this compositional change does not have to be linear as exponential, sigmoidal and radial gradients have been fabricated [65–67]. The growing inclusion of orthogonal chemistries is increasing the number of test parameter gradients, which can be overlaid in a single sample [68,69]. Due to this flexibility, variations in fabrication of gradient samples have ranged from complex microfluidics, which can directly overlay two orthogonal gradients for the study of multiple parameters at a time [70], to large gradients (6 cm by 6 cm) fabricated using a peristaltic pump drawing from two polymer reservoirs into a mold in order to create many replicates or large samples for complex analysis methods from the same gradient hydrogel [71,72].

**Figure 1.** Demonstration of colormetric gradient in a polyethylene glycol hydrogel. Adapted with the permission from [73]. Copyright 2015 Elsevier.

Like arrays, gradient samples have been used to characterize changes in material properties due alterations in polymer chemistry and hydrogel formulation [74,75]. These characterization studies have been expanded to examine the linear and non-linear gradient release of growth factors and drugs from hydrogels in order to optimize release kinetics from the hydrogel for localized delivery [76–79]. Even short interfering RNA gradients have been fabricated [80], with one study demonstrating complete silencing of green fluorescent protein expression in encapsulated cells with high concentrations of incorporated short interfering RNA in the hydrogel [80].

Due to the continuously changing nature of the gradient samples, the number of distinct formulations tested cannot be directly calculated like in DOE and arrays formats. As such, isolating the optimal composition after testing can be more difficult in gradient samples than DOE and array samples. Good characterization of the gradient's material properties is necessary for identification of the optimal test conditions to occur. It also identifies confounding factors due to material property changes in the sample other than the test parameter along the length of the gradient created due to fabrication of the gradient for the test parameter. This is important because changes in bioactive signaling inclusion and polymer composition have been shown to alter a wide range of material properties in hydrogels such as Young's modulus, mesh size and swelling ratio [81–83]. Often DOE and array samples rely on formulation data to identify the optimized condition, instead of directly measuring material properties in the fabricated samples. As all of the included components may not have anchored into the hydrogel during fabrication and additional material properties may have been altered beyond the test parameter, this may not be an accurate presentation of the hydrogel environment the cells interacted with. A strategy based on direct measurement of all test and material properties in the fabricated hydrogel system will provide the most accurate determination of the optimized condition with all combinatorial method/high throughput approaches, not just gradient samples.

The complex relationship between hydrogel material properties and cellular response makes this level of characterization even more critical. Small changes in a number of factors, even some of which were unintended, may alter the observed cellular response. To illustrate this, changes in hydrogel thickness, fiber density and stiffness have been observed along polymer concentration gradients [83–85], each of which can alter cellular response to the hydrogel [5,86,87]. Overlapping gradients in hydrogel wettability and stiffness found an interaction between the two material properties affecting hMSC adhesion and spreading, where changes in wettability lead to alterations in the material stiffness where hMSC adhesion and spreading occurred [88]. This study demonstrates the complexity of these interactions between multiple material properties on cellular behavior. Compositional blending gradients of polymers or solvents at different ratios have been used to form porosity gradients in hydrogels [85,89,90]. Changes in porosity across the gradient have been found to alter cell cytoskeletal structure [90], which can alter later differentiation [91]. It is important to note that one study demonstrated a difference in hydrogel stiffness along with the porosity change [85], which would complicate analysis of cellular results due to the porosity change. Reaction kinetics in hydrogel photopolymerization based on ultraviolet light exposure time have been monitored [92]. Low conversion rates due to insufficient ultraviolet light exposure were associated with reduced macrophage viability and increased expression of inflammation markers by the cells [64,93]. This demonstrates that not just formulation changes, but also the efficiency of the system to completely consume the reactive elements utilized for gelation or functionalization can have significant effect on cellular response.

One of the major advantages of the gradient approach over DOE and arrays in biological studies is that every possible concentration or combination of test parameters within the test range is present in the gradient hydrogel and is routinely examined. This makes the approach particularly well suited for the optimization of the test parameters. The small changes in sample composition across gradient samples have been found to affect cellular attachment [90,92,94–97], viability [27,64,72], migration [66,69,70,98–100] and differentiation [71–73,83,101–104]. To demonstrate this phenomena, Figure 2 shows a study of the effects of a Young's modulus gradient on human chondrocyte glycosaminoglycan content. Detection of these shifts in cellular behavior are easier in gradient samples than in DOE and arrays due to reduced sample preparation, which limits sample variation effect on the biological results.

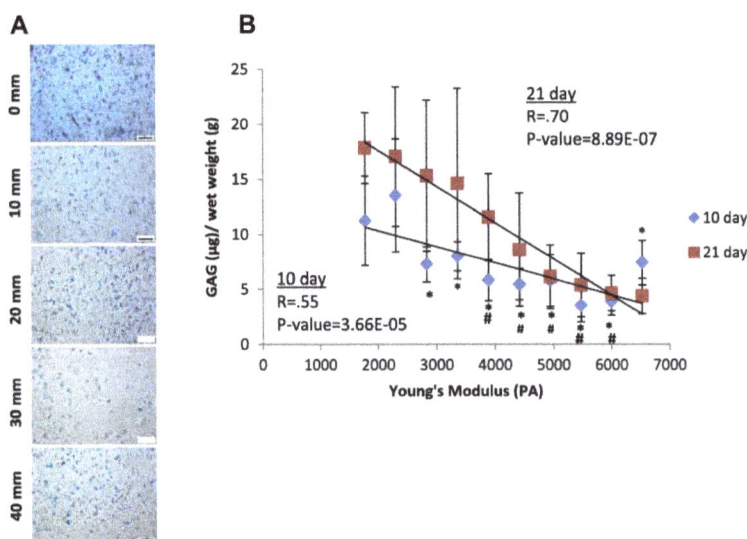

**Figure 2.** Extracellular matrix (ECM) production by human chondrocytes after 10 days of culture. Images were taken at 10 mm intervals along the length of the modulus gradient. (**A**) Whole mount Alcian blue; (**B**) sulfated gylcosaminoglycan quantification based on Alcian blue extraction shows distinct changes with position in the modulus profile at both 10 and 21 days. Scale bar 200 μm. # indicates $p \leqslant 0.05$ compared with the 1700 Pa Young's modulus gradient position; * indicates $p \leqslant 0.05$ compared with the 2300 Pa Young's modulus gradient position. R (coefficient of multiple correlation) and P-value indicate statistical results of linear regression analysis, and indicate a high confidence in the linear relationship in the data. Adapted with the permission from [83]. Copyright 2013 Elsevier.

Due to the ease and flexibility of fabrication, a number of studies have utilized gradient hydrogel systems to examine the effect of material property changes spanning from studies of Young's modulus [11,66,70,83,92,96,97,99,101,104–106] to bioactive signal concentration [19,22,76–79,90,94,95,100,102,107] on cellular behavior with numerous cell types. Large changes in hydrogel stiffness have been shown to affect the lineage choice of stem cells and cellular proliferation [108]. Small changes have been shown to affect cellular function, as in Figure 2, where ECM content was altered [83]. Both small and large stiffness gradients have demonstrated changes in cytoskeletal structure [62,83]. Similar changes in cytoskeletal structure have been observed in bioactive signal concentration gradients [102]. These early changes in cellular behavior due to interactions with the hydrogel could led to the differences in cell response observed at later time points [91].

Ideally, the gradients should not be sensed at the cellular level and cells respond as if in a homogenous material. However, steep gradients, which were sensed by individual cells have been observed and led to the cellular alignment along the gradient [100,107,109–111]. There are certain biological systems where this alignment is advantageous, for instance when used to develop a predictive model of cell migration [112], direct cellular migration down a mechanotactic gradient [61], or recapitulate a biological gradient to aid in tissue formation [113,114]. However, it is often an unintended confounding factor, which complicates the material optimization process for tissue engineering. Once identified the test range for the given parameter can easily be altered to eliminate cellular alignment along the gradient. Shifting the test parameter range within gradient samples does not require many changes to sample fabrication. The technique of altering the test parameter range has, also, been used to study regions of interest at greater resolution [73]. This allows for fine tuning of the optimized condition to maximize the desired cellular effects quickly.

Complex biological systems can be studied with a gradient approach using overlapping or sequential gradients [84,92,115]. Using overlapping gradients of nerve growth factor and neurotrophin-3 immobilized in poly(2-hydroxyethylmethacrylate) hydrogels, one study identified a synergistic, and not merely additive, effect of the proteins on chick dorsal root ganglia neurite extension [76]. A study of orthogonal Young's Modulus and protein concentration gradients found cellular migration distance and velocity increased with increasing hydrogel Young's modulus at low hepatocyte growth factor concentration, but that changes in Young's modulus had no effect on migration distance and velocity at a high hepatocyte growth factor concentrations [70]. Again demonstrating the complexity of relationships among multiple biological signals, and how under the right conditions one can play a dominate role over the others in directing cellular response. Another study was able to provide real time monitoring of endothelial cells while controlling hydrogel properties, solute gradients, surface shear stress and interstitial flow through the matrix [98], helping to determine the optimal combination to spur vessel formation. Beyond high throughput analysis, these systems can be further developed to emulate tissue development and native function. These *in vitro* models can then serve as drug testing platforms or models to study human development and disease progression.

## 5. Combinatorial Method/High Throughput Sample Design Considerations

Regardless of the combinatorial method/high throughput approach used to conduct these studies, there are number of design parameters that should be considered when cell culture experiments are being conducted. The first is that there are limitations on the selection of polymers and gelation approaches, which can be utilized in these approaches. Inappropriate viscosity and gelation kinetics can lead to inconsistent sample formation. This is particularly true for three-dimensional culture systems as the cellular distribution may not be consistent throughout the sample, potentially altering cellular behavior in the sample and experimental results. Due to its robust consistent network formation and speed [66,73,93], photopolymerization has most often been used for gelation of combinatorial method/high throughput samples. However, reduction-oxidative reactions have proven suitable in certain sample fabrications for use in combinatorial method/high throughput approaches [31,41].

Even if cellular distribution is homogenous across the sample, major differences in the cellular environment exist between two-and three-dimensional culture that can alter the optimal hydrogel formulation [116]. This was effectively demonstrated in a recent study examining the effects of IKVAV concentration on mouse embryonic stem cell neural differentiation [73]. Not only was a significant drop in the IKVAV concentration capable of promoting neurite extension observed in three-dimensional culture compared to two-dimensional culture, but also a significant delay in neural differentiation in three-dimensional culture compared to two-dimensional culture (Figure 3). This study demonstrates that extrapolation of optimal conditions from two- to three-dimensional studies will be at best difficult. As both culture methods are useful for cellular expansion and tissue formations in tissue engineering applications, both culture methods need to continue to be studied with a systematic approach.

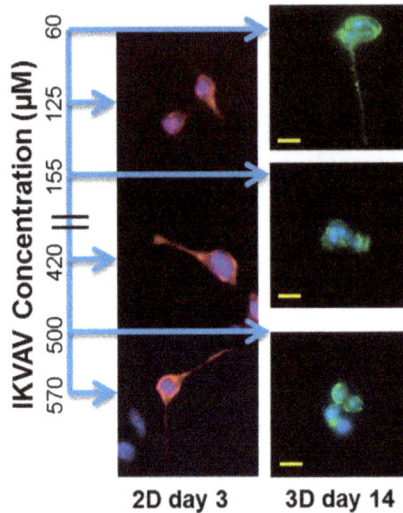

**Figure 3.** Beta III tubulin staining of neurite extension of cells in 2D (red) and 3D (green) culture with nuclear staining (blue) exposed to a continuous IKVAV gradient in polyethylene glycol hydrogels after 3 days of culture in 2D and 14 days of culture in 3D. Scale bars: 10 μm. Adapted with permission from [73]. Copyright 2015 Elsevier.

The effects of cellular crosstalk between positions is another critical design consideration in combinatorial method/high throughput systems. hMSC lineage choice was used to examine the effects of cellular crosstalk in a recent study [103]. Access to cell secreted cytokines was either freely allowed across an RGD concentration gradient hydrogel or restricted through sectioning and discrete culture of gradient sections in isolated tissue culture wells [103]. The study found that free access to cytokines from hMSC exposed to all test RGD concentrations favored adipogenic differentiation, while restricted access to only cytokines secreted from hMSC exposed to similar RGD concentrations favored osteogenic differentiation (Figure 4) [103]. This highlights to potential effect of cell secreted cytokines as a confounding factor in analysis of combinatorial method/high throughput systems, which can alter biological results. The effects of crosstalk can become even more complicated when more than one cell type is included in the combinatorial/ high throughput sample [40,48,117].

As technology advances and designs of combinatorial method/high throughput samples become more complex, isolating the effects of each test parameter on cellular response will be increasing important. As shown by the complex studies already in the literature, one component can dominate the cellular response masking the effects of the others if not properly managed [70]. Alterations in one test parameter can also modulate the response of another, providing apparently conflicting results in terms of biological response for the second parameter [88]. Expansion of temporal studies, pose additional complexity as the cellular differentiation state and even point in cell cycle can alter response. These technological advances will likely lead to utilization of more advanced biological outputs in combinatorial method/high throughput strategies. As the complexity of the biological outputs increases, the chances of identify more interconnected test parameters will also increase.

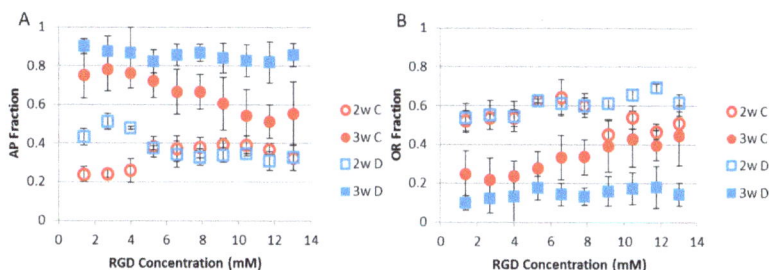

**Figure 4.** Effect of gradient culture condition on human mesenchymal stem cell lineage selection. (**A**) Fraction of cells expressing alkaline phosphatase, an osteogenic marker; (**B**) Fraction of cells with adipogenic vacuole staining in continuous culture, (**C**) which allows free access to cytokines secreted from cells across the gradient, and discrete culture, (**D**) which limits cytokine access to those secreted from cells in nearly similar RGD concentrations. Adapted and Reprinted with permission from [103]. Copyright 2013 American Chemical Society.

## 6. Concluding Remarks

Broader adoption of these combinatorial method/high throughput strategies is necessary is to efficiently optimize hydrogels and bring the promise of tissue engineering closer to fruition. The extent of how hydrogels influence cellular behavior has not yet been fully elucidated. Alterations in cellular response due to cellular differentiation states, species of cellular origin and culture type are just beginning to be understood [10,73,118–120]. Combinatorial method/high throughput strategies offer the ability to systematically develop the base of knowledge necessary for model development, which will finally enable the rational design of biomaterials to emulate key ECM factors governing cellular behavior. However, as the biology of cell-material interaction becomes better appreciated and the complexity of biological outputs advances with technology in combinatorial method/high throughput strategies, so must the level of material characterization. Post-fabrication, as well as during culture, material characterization must occur in order to keep pace with the biology in order to truly elucidate how the two systems (material and biological) alter each other over time. This will require the development of new material characterization procedures.

DOE, arrays and gradient samples, the three methods discussed in this review that have already been applied to hydrogel optimization, are just the first step in this movement toward systematic studies and rational design. Each has its own advantages and drawbacks when utilized in the optimization process, which makes it better suited for particular types of studies. However, they all allow for the optimization of multiple test parameters simultaneously using a systematic approach, which has been shown in each case to identify parameter interactions that traditional methods of hydrogel optimization have failed to identify. As the field advances, these strategies as well as new ones which have not been developed or adapted to hydrogel development will be utilized from discovery to final optimization in combination. They will become even more powerful tools for hydrogel design and optimization. Imagine the additional efficiency obtained from using DOE to dictate sample formulations in a miniaturized array. The results from those experiments could then be fine-tuned in gradient samples to obtain the final optimization. In fact, the cross utilization of multiple combinatorial method/high throughput strategies in a single study has already begun [108,121]. This should decrease everything from total development time to research cost. A move that will bring the hydrogel development process much more in line with the pharmaceutical industry and hopefully bring many more tissue engineering based treatments to the clinical application with greater speed.

**Acknowledgments:** The author would like to acknowledge financial support from the following sources: Mission Connect, a TIRR program (014-120), The Staman Ogilvie Fund, William Stamps Farish Fund, Bentsen Stroke Center, and Vivian L. Smith Department of Neurosurgery.

**Conflicts of Interest:** The author declares no conflict of interest.

## Abbreviations

The following abbreviations are used in this manuscript:

| | |
|---|---|
| ECM | extracellular matrix |
| DOE | design of experiment |
| hMSC | human mesenchymal stem cells |
| HUVEC | human umbilical vein endothelial cell |
| hESC | human embryonic stem cell |

## References

1. Alberts, B.; Johnson, A.; Lewis, J. The extracellular matrix of animals. In *Molecular Biology of the Cell*, 4th ed.; Graland Science: New York, NY, USA, 2002.
2. Costa, P.; Almeida, F.V.M.; Connelly, J.T. Biophysical signals controlling cell fate decisions: How do stem cells really feel? *Int. J. Biochem. Cell Biol.* **2012**, *44*, 2233–2237. [CrossRef] [PubMed]
3. Van Dijk, M.; Göransson, S.A.; Strömblad, S. Cell to extracellular matrix interactions and their reciprocal nature in cancer. *Exp. Cell Res.* **2013**, *319*, 1663–1670. [CrossRef] [PubMed]
4. Smith, L.A.; Liu, X.; Hu, J.; Ma, P.X. The influence of three-dimensional nanofibrous scaffolds on the osteogenic differentiation of embryonic stem cells. *Biomaterials* **2009**, *30*, 2516–2522. [CrossRef] [PubMed]
5. Engler, A.J.; Sen, S.; Sweeney, H.L.; Discher, D.E. Matrix elasticity directs stem cell lineage specification. *Cell* **2006**, *126*, 677–689. [CrossRef] [PubMed]
6. Smith, L.A.; Liu, X.; Ma, P.X. Tissue engineering with nano-fibrous scaffolds. *Soft Matter* **2008**, *4*, 2144–2149. [CrossRef] [PubMed]
7. Vats, K.; Benoit, D.S.W. Dynamic manipulation of hydrogels to control cell behavior: A review. *Tissue Eng. Part B Rev.* **2013**, *19*, 455–469. [CrossRef] [PubMed]
8. Janson, I.A.; Putnam, A.J. Extracellular matrix elasticity and topography: Material-based cues that affect cell function via conserved mechanisms. *J. Biomed. Mater. Res. A* **2015**, *103*, 1246–1258. [CrossRef] [PubMed]
9. Buwalda, S.J.; Boere, K.W.M.; Dijkstra, P.J.; Feijen, J.; Vermonden, T.; Hennink, W.E. Hydrogels in a historical perspective: From simple networks to smart materials. *J. Control. Release* **2014**, *190*, 254–273. [CrossRef] [PubMed]
10. Callahan, L.A.S.; Ganios, A.M.; McBurney, D.L.; Dilisio, M.F.; Weiner, S.D.; Horton, W.E.; Becker, M.L. Ecm production of primary human and bovine chondrocytes in hybrid peg hydrogels containing type i collagen and hyaluronic acid. *Biomacromolecules* **2012**, *13*, 1625–1631. [CrossRef] [PubMed]
11. Zheng, J.; Smith Callahan, L.A.; Hao, J.; Guo, K.; Wesdemiotis, C.; Weiss, R.A.; Becker, M.L. Strain-promoted cross-linking of peg-based hydrogels via copper- free cycloaddition. *ACS Macro Lett.* **2012**, *1*, 1071–1073. [CrossRef] [PubMed]
12. Bian, L.; Guvendiren, M.; Mauck, R.L.; Burdick, J.A. Hydrogels that mimic developmentally relevant matrix and n-cadherin interactions enhance msc chondrogenesis. *Proc. Natl. Acad. Sci. USA* **2013**, *110*, 10117–10122. [CrossRef] [PubMed]
13. Tang, S.; Glassman, M.J.; Li, S.; Socrate, S.; Olsen, B.D. Oxidatively responsive chain extension to entangle engineered protein hydrogels. *Macromolecules* **2014**, *47*, 791–799. [CrossRef] [PubMed]
14. Liu, Z.; Lin, Q.; Sun, Y.; Liu, T.; Bao, C.; Li, F.; Zhu, L. Spatiotemporally controllable and cytocompatible approach builds 3d cell culture matrix by photo-uncaged-thiol michael addition reaction. *Adv. Mater.* **2014**, *26*, 3912–3917. [CrossRef] [PubMed]
15. Pritchard, C.D.; O'Shea, T.M.; Siegwart, D.J.; Calo, E.; Anderson, D.G.; Reynolds, F.M.; Thomas, J.A.; Slotkin, J.R.; Woodard, E.J.; Langer, R. An injectable thiol-acrylate poly(ethylene glycol) hydrogel for sustained release of methylprednisolone sodium succinate. *Biomaterials* **2011**, *32*, 587–597. [CrossRef] [PubMed]

16. Zisch, A.H.; Lutolf, M.P.; Ehrbar, M.; Raeber, G.P.; Rizzi, S.C.; Davies, N.; Schmökel, H.; Bezuidenhout, D.; Djonov, V.; Zilla, P.; *et al.* Cell-demanded release of VEGF from synthetic, biointeractive cell-ingrowth matrices for vascularized tissue growth. *FASEB J.* **2003**, *17*, 2260–2262. [CrossRef] [PubMed]

17. Singh, S.P.; Schwartz, M.P.; Tokuda, E.Y.; Luo, Y.; Rogers, R.E.; Fujita, M.; Ahn, N.G.; Anseth, K.S. A synthetic modular approach for modeling the role of the 3d microenvironment in tumor progression. *Sci. Rep.* **2015**, *5*, 17814. [CrossRef] [PubMed]

18. Guvendiren, M.; Burdick, J.A. Engineering synthetic hydrogel microenvironments to instruct stem cells. *Curr. Opin. Biotechnol.* **2013**, *24*, 841–846. [CrossRef] [PubMed]

19. Schweller, R.M.; West, J.L. Encoding hydrogel mechanics via network cross-linking structure. *ACS Biomater. Sci. Eng.* **2015**, *1*, 335–344. [CrossRef] [PubMed]

20. LaNasa, S.M.; Hoffecker, I.T.; Bryant, S.J. Presence of pores and hydrogel composition influence tensile properties of scaffolds fabricated from well-defined sphere templates. *J. Biomed. Mater. Res. B Appl. Biomater.* **2011**, *96B*, 294–302. [CrossRef] [PubMed]

21. Hook, A.L.; Anderson, D.G.; Langer, R.; Williams, P.; Davies, M.C.; Alexander, M.R. High throughput methods applied in biomaterial development and discovery. *Biomaterials* **2010**, *31*, 187–198. [CrossRef] [PubMed]

22. Anderson, D.G.; Levenberg, S.; Langer, R. Nanoliter-scale synthesis of arrayed biomaterials and application to human embryonic stem cells. *Nat. Biotechnol.* **2004**, *22*, 863–866. [CrossRef] [PubMed]

23. Smith Callahan, L.A.; Ma, Y.; Stafford, C.M.; Becker, M.L. Concentration dependent neural differentiation and neurite extension of mouse esc on primary amine-derivatized surfaces. *Biomater. Sci.* **2013**, *1*, 537–544. [CrossRef]

24. Kim, H.D.; Lee, E.A.; Choi, Y.H.; An, Y.H.; Koh, R.H.; Kim, S.L.; Hwang, N.S. High throughput approaches for controlled stem cell differentiation. *Acta Biomater.* **2016**, *34*, 21–29. [CrossRef] [PubMed]

25. Oliveira, M.B.; Mano, J.F. High-throughput screening for integrative biomaterials design: Exploring advances and new trends. *Trends Biotechnol.* **2014**, *32*, 627–636. [CrossRef] [PubMed]

26. Chen, X.C.; Zhou, L.; Gupta, S.; Civoli, F. Implementation of design of experiments (DOE) in the development and validation of a cell-based bioassay for the detection of anti-drug neutralizing antibodies in human serum. *J. Immunol. Methods* **2012**, *376*, 32–45. [CrossRef] [PubMed]

27. Bersini, S.; Gilardi, M.; Arrigoni, C.; Talo, G.; Zamai, M.; Zagra, L.; Caiolfa, V.; Moretti, M. Human *in vitro* 3d co-culture model to engineer vascularized bone-mimicking tissues combining computational tools and statistical experimental approach. *Biomaterials* **2016**, *76*, 157–172. [CrossRef] [PubMed]

28. Jung, J.P.; Moyano, J.V.; Collier, J.H. Multifactorial optimization of endothelial cell growth using modular synthetic extracellular matrices. *Integr. Biol.* **2011**, *3*, 185–196. [CrossRef] [PubMed]

29. Lam, J.; Carmichael, S.T.; Lowry, W.E.; Segura, T. Hydrogel design of experiments methodology to optimize hydrogel for iPSC-NPC culture. *Adv. Healthc. Mater.* **2015**, *4*, 534–539. [CrossRef] [PubMed]

30. Lee, J.; Abdeen, A.A.; Zhang, D.; Kilian, K.A. Directing stem cell fate on hydrogel substrates by controlling cell geometry, matrix mechanics and adhesion ligand composition. *Biomaterials* **2013**, *34*, 8140–8148. [CrossRef] [PubMed]

31. Zhang, R.; Liberski, A.; Khan, F.; Diaz-Mochon, J.J.; Bradley, M. Inkjet fabrication of hydrogel microarrays using *in situ* nanolitre-scale polymerisation. *Chem. Commun.* **2008**, 1317–1319. [CrossRef] [PubMed]

32. Koh, W.-G.; Itle, L.J.; Pishko, M.V. Molding of hydrogel microstructures to create multiphenotype cell microarrays. *Anal. Chem.* **2003**, *75*, 5783–5789. [CrossRef] [PubMed]

33. Jongpaiboonkit, L.; King, W.J.; Lyons, G.E.; Paguirigan, A.L.; Warrick, J.W.; Beebe, D.J.; Murphy, W.L. An adaptable hydrogel array format for 3-dimensional cell culture and analysis. *Biomaterials* **2008**, *29*, 3346–3356. [CrossRef] [PubMed]

34. King, W.J.; Jongpaiboonkit, L.; Murphy, W.L. Influence of FGF2 and PEG hydrogel matrix properties on hmsc viability and spreading. *J. Biomed. Mater. Res. A* **2010**, *93*, 1110–1123. [CrossRef] [PubMed]

35. Le, N.N.; Zorn, S.; Schmitt, S.K.; Gopalan, P.; Murphy, W.L. Hydrogel arrays formed via differential wettability patterning enable combinatorial screening of stem cell behavior. *Acta Biomater.* **2015**. [CrossRef] [PubMed]

36. Dolatshahi-Pirouz, A.; Nikkhah, M.; Gaharwar, A.K.; Hashmi, B.; Guermani, E.; Aliabadi, H.; Camci-Unal, G.; Ferrante, T.; Foss, M.; Ingber, D.E.; *et al.* A combinatorial cell-laden gel microarray for inducing osteogenic differentiation of human mesenchymal stem cells. *Sci. Rep.* **2014**, *4*, 3896. [CrossRef] [PubMed]

37. Ranga, A.; Gobaa, S.; Okawa, Y.; Mosiewicz, K.; Negro, A.; Lutolf, M.P. 3D niche microarrays for systems-level analyses of cell fate. *Nat. Commun.* **2014**, *5*. [CrossRef] [PubMed]

38. Neuss, S.; Apel, C.; Buttler, P.; Denecke, B.; Dhanasingh, A.; Ding, X.; Grafahrend, D.; Groger, A.; Hemmrich, K.; Herr, A.; *et al.* Assessment of stem cell/biomaterial combinations for stem cell-based tissue engineering. *Biomaterials* **2008**, *29*, 302–313. [CrossRef] [PubMed]

39. Eun Chung, S.; Kim, J.; Yoon Oh, D.; Song, Y.; Lee, S.H.; Min, S.; Kwon, S. One-step pipetting and assembly of encoded chemical-laden microparticles for high-throughput multiplexed bioassays. *Nat. Commun.* **2014**, *5*. [CrossRef] [PubMed]

40. Li, Y.; Chen, P.; Wang, Y.; Yan, S.; Feng, X.; Du, W.; Koehler, S.A.; Demirci, U.; Liu, B.-F. Rapid assembly of heterogeneous 3D cell microenvironments in a microgel array. *Adv. Mater.* **2016**, *28*, 3543–3548. [CrossRef] [PubMed]

41. Zhang, R.; Liberski, A.; Sanchez-Martin, R.; Bradley, M. Microarrays of over 2000 hydrogels—Identification of substrates for cellular trapping and thermally triggered release. *Biomaterials* **2009**, *30*, 6193–6201. [CrossRef] [PubMed]

42. Mei, Y.; Gerecht, S.; Taylor, M.; Urquhart, A.J.; Bogatyrev, S.R.; Cho, S.-W.; Davies, M.C.; Alexander, M.R.; Langer, R.S.; Anderson, D.G. Mapping the interactions among biomaterials, adsorbed proteins, and human embryonic stem cells. *Adv. Mater.* **2009**, *21*, 2781–2786. [CrossRef]

43. Neto, A.I.; Correia, C.R.; Custódio, C.A.; Mano, J.F. Biomimetic miniaturized platform able to sustain arrays of liquid droplets for high-throughput combinatorial tests. *Adv. Funct. Mater.* **2014**, *24*, 5096–5103. [CrossRef]

44. Yang, J.; Mei, Y.; Hook, A.L.; Taylor, M.; Urquhart, A.J.; Bogatyrev, S.R.; Langer, R.; Anderson, D.G.; Davies, M.C.; Alexander, M.R. Polymer surface functionalities that control human embryoid body cell adhesion revealed by high throughput surface characterization of combinatorial material microarrays. *Biomaterials* **2010**, *31*, 8827–8838. [CrossRef] [PubMed]

45. Unciti-Broceta, A.; Díaz-Mochón, J.J.; Mizomoto, H.; Bradley, M. Combining nebulization-mediated transfection and polymer microarrays for the rapid determination of optimal transfection substrates. *J. Comb. Chem.* **2008**, *10*, 179–184. [CrossRef] [PubMed]

46. Tang, M.D.; Golden, A.P.; Tien, J. Fabrication of collagen gels that contain patterned, micrometer-scale cavities. *Adv. Mater.* **2004**, *16*, 1345–1348. [CrossRef]

47. Floren, M.; Tan, W. Three-dimensional, soft neotissue arrays as high throughput platforms for the interrogation of engineered tissue environments. *Biomaterials* **2015**, *59*, 39–52. [CrossRef] [PubMed]

48. Albrecht, D.R.; Tsang, V.L.; Sah, R.L.; Bhatia, S.N. Photo- and electropatterning of hydrogel-encapsulated living cell arrays. *Lab Chip* **2005**, *5*, 111–118. [CrossRef] [PubMed]

49. Patel, R.G.; Purwada, A.; Cerchietti, L.; Inghirami, G.; Melnick, A.; Gaharwar, A.K.; Singh, A. Microscale bioadhesive hydrogel arrays for cell engineering applications. *Cell. Mol. Bioeng.* **2014**, *7*, 394–408. [CrossRef] [PubMed]

50. Duffy, C.; Venturato, A.; Callanan, A.; Lilienkampf, A.; Bradley, M. Arrays of 3d double-network hydrogels for the high-throughput discovery of materials with enhanced physical and biological properties. *Acta Biomater.* **2015**, *34*, 104–112. [CrossRef] [PubMed]

51. Kurkuri, M.D.; Driever, C.; Johnson, G.; McFarland, G.; Thissen, H.; Voelcker, N.H. Multifunctional polymer coatings for cell microarray applications. *Biomacromolecules* **2009**, *10*, 1163–1172. [CrossRef] [PubMed]

52. Hansen, T.D.; Koepsel, J.T.; Le, N.N.; Nguyen, E.H.; Zorn, S.; Parlato, M.; Loveland, S.G.; Schwartz, M.P.; Murphy, W.L. Biomaterial arrays with defined adhesion ligand densities and matrix stiffness identify distinct phenotypes for tumorigenic and nontumorigenic human mesenchymal cell types. *Biomater. Sci.* **2014**, *2*, 745–756. [CrossRef] [PubMed]

53. Jongpaiboonkit, L.; King, W.J.; Murphy, W.L. Screening for 3D environments that support human mesenchymal stem cell viability using hydrogel arrays. *Tissue Eng. A* **2009**, *15*, 343–353. [CrossRef] [PubMed]

54. Gobaa, S.; Hoehnel, S.; Roccio, M.; Negro, A.; Kobel, S.; Lutolf, M.P. Artificial niche microarrays for probing single stem cell fate in high throughput. *Nat. Methods* **2011**, *8*, 949–955. [CrossRef] [PubMed]

55. Nguyen, E.H.; Zanotelli, M.R.; Schwartz, M.P.; Murphy, W.L. Differential effects of cell adhesion, modulus and vegfr-2 inhibition on capillary network formation in synthetic hydrogel arrays. *Biomaterials* **2014**, *35*, 2149–2161. [CrossRef] [PubMed]

56. Thissen, H.; Johnson, G.; McFarland, G.; Verbiest, B.C.H.; Gengenbach, T.; Voelcker, N.H. Microarrays for the Evaluation of Cell-Biomaterial Surface Interactions. *Proc. SPIE 6413*, Smart Materials IV. 64130B.

57. Mant, A.; Tourniaire, G.; Diaz-Mochon, J.J.; Elliott, T.J.; Williams, A.P.; Bradley, M. Polymer microarrays: Identification of substrates for phagocytosis assays. *Biomaterials* **2006**, *27*, 5299–5306. [CrossRef] [PubMed]

58. Ueda, E.; Geyer, F.L.; Nedashkivska, V.; Levkin, P.A. Dropletmicroarray: Facile formation of arrays of microdroplets and hydrogel micropads for cell screening applications. *Lab Chip* **2012**, *12*, 5218–5224. [CrossRef] [PubMed]

59. Sant, S.; Hancock, M.J.; Donnelly, J.P.; Iyer, D.; Khademhosseini, A. Biomimetic gradient hydrogels for tissue engineering. *Can. J. Chem. Eng.* **2010**, *88*, 899–911. [CrossRef] [PubMed]

60. Meyvantsson, I.; Warrick, J.W.; Hayes, S.; Skoien, A.; Beebe, D.J. Automated cell culture in high density tubeless microfluidic device arrays. *Lab Chip* **2008**, *8*, 717–724. [CrossRef] [PubMed]

61. Cai, P.; Layani, M.; Leow, W.R.; Amini, S.; Liu, Z.; Qi, D.; Hu, B.; Wu, Y.-L.; Miserez, A.; Magdassi, S.; *et al.* Bio-inspired mechanotactic hybrids for orchestrating traction-mediated epithelial migration. *Adv. Mater.* **2016**, *28*, 3102–3110. [CrossRef] [PubMed]

62. Kim, T.H.; An, D.B.; Oh, S.H.; Kang, M.K.; Song, H.H.; Lee, J.H. Creating stiffness gradient polyvinyl alcohol hydrogel using a simple gradual freezing–thawing method to investigate stem cell differentiation behaviors. *Biomaterials* **2015**, *40*, 51–60. [CrossRef] [PubMed]

63. Yi, Z.; Zhang, Y.; Kootala, S.; Hilborn, J.; Ossipov, D.A. Hydrogel patterning by diffusion through the matrix and subsequent light-triggered chemical immobilization. *ACS Appl. Mater. Interfaces* **2015**, *7*, 1194–1206. [CrossRef] [PubMed]

64. Lin, N.J.; Drzal, P.L.; Lin-Gibson, S. Two-dimensional gradient platforms for rapid assessment of dental polymers: A chemical, mechanical and biological evaluation. *Dent. Mater.* **2007**, *23*, 1211–1220. [CrossRef] [PubMed]

65. Selimović, Š.; Sim, W.Y.; Kim, S.B.; Jang, Y.H.; Lee, W.G.; Khabiry, M.; Bae, H.; Jambovane, S.; Hong, J.W.; Khademhosseini, A. Generating nonlinear concentration gradients in microfluidic devices for cell studies. *Anal. Chem.* **2011**, *83*, 2020–2028. [CrossRef] [PubMed]

66. Wong, J.Y.; Velasco, A.; Rajagopalan, P.; Pham, Q. Directed movement of vascular smooth muscle cells on gradient-compliant hydrogels. *Langmuir* **2003**, *19*, 1908–1913. [CrossRef]

67. Lin, F.; Saadi, W.; Rhee, S.W.; Wang, S.-J.; Mittal, S.; Jeon, N.L. Generation of dynamic temporal and spatial concentration gradients using microfluidic devices. *Lab Chip* **2004**, *4*, 164–167. [CrossRef] [PubMed]

68. Allazetta, S.; Cosson, S.; Lutolf, M.P. Programmable microfluidic patterning of protein gradients on hydrogels. *Chem. Commun.* **2011**, *47*, 191–193. [CrossRef] [PubMed]

69. Cosson, S. Capturing complex protein gradients on biomimetic hydrogels for cell-based assays. *Adv. Funct. Mater.* **2009**, *19*, 3411–3419. [CrossRef]

70. Garcia, S.; Sunyer, R.; Olivares, A.; Noailly, J.; Atencia, J.; Trepat, X. Generation of stable orthogonal gradients of chemical concentration and substrate stiffness in a microfluidic device. *Lab Chip* **2015**, *15*, 2606–2614. [CrossRef] [PubMed]

71. Chatterjee, K.; Lin-Gibson, S.; Wallace, W.E.; Parekh, S.H.; Lee, Y.J.; Cicerone, M.T.; Young, M.F.; Simon, C.G., Jr. The effect of 3d hydrogel scaffold modulus on osteoblast differentiation and mineralization revealed by combinatorial screening. *Biomaterials* **2010**, *31*, 5051–5062. [CrossRef] [PubMed]

72. Chatterjee, K.; Young, M.F.; Simon, C.G. Fabricating gradient hydrogel scaffolds for 3D cell culture. *Comb. Chem. High Throughput Screen.* **2011**, *14*, 227–236. [CrossRef] [PubMed]

73. Yang, Y.H.; Khan, Z.; Ma, C.; Lim, H.J.; Smith Callahan, L.A. Optimization of adhesive conditions for neural differentiation of murine embryonic stem cells using hydrogels functionalized with continuous ile-LYs-Val-Ala-Val concentration gradients. *Acta Biomater.* **2015**, *21*, 55–62. [CrossRef] [PubMed]

74. Johnson, P.M.; Reynolds, T.B.; Stansbury, J.W.; Bowman, C.N. High throughput kinetic analysis of photopolymer conversion using composition and exposure time gradients. *Polymer* **2005**, *46*, 3300–3306. [CrossRef]

75. Bailey, B.M.; Nail, L.N.; Grunlan, M.A. Continuous gradient scaffolds for rapid screening of cell–material interactions and interfacial tissue regeneration. *Acta Biomater.* **2013**, *9*, 8254–8261. [CrossRef] [PubMed]

76. Moore, K.; MacSween, M.; Shoichet, M. Immobilized concentration gradients of neurotrophic factors guide neurite outgrowth of primary neurons in macroporous scaffolds. *Tissue Eng.* **2006**, *12*, 267–278. [CrossRef] [PubMed]

77. Peret, B.J.; Murphy, W.L. Controllable soluble protein concentration gradients in hydrogel networks. *Adv. Funct. Mater.* **2008**, *18*, 3410–3417. [CrossRef] [PubMed]

78. Lee, P.I. Effect of non-uniform initial drug concentration distribution on the kinetics of drug release from glassy hydrogel matrices. *Polymer* **1984**, *25*, 973–978. [CrossRef]

79. Wang, X.; Wenk, E.; Zhang, X.; Meinel, L.; Vunjak-Novakovic, G.; Kaplan, D.L. Growth factor gradients via microsphere delivery in biopolymer scaffolds for osteochondral tissue engineering. *J. Control. Release Off. J. Control. Release Soc.* **2009**, *134*, 81–90. [CrossRef] [PubMed]

80. Hill, M.C.; Nguyen, M.K.; Jeon, O.; Alsberg, E. Spatial control of cell gene expression by sirna gradients in biodegradable hydrogels. *Adv. Healthc. Mater.* **2015**, *4*, 714–722. [CrossRef] [PubMed]

81. Villanueva, I.; Weigel, C.A.; Bryant, S.J. Cell-matrix interactions and dynamic mechanical loading influence chondrocyte gene expression and bioactivity in peg-rgd hydrogels. *Acta Biomater.* **2009**, *5*, 2832–2846. [CrossRef] [PubMed]

82. Zustiak, S.P.; Durbal, R.; Leach, J.B. Influence of cell-adhesive peptide ligands on poly(ethylene glycol) hydrogel physical, mechanical and transport properties. *Acta Biomater.* **2010**, *6*, 3404–3414. [CrossRef] [PubMed]

83. Smith Callahan, L.A.; Ganios, A.M.; Childers, E.P.; Weiner, S.D.; Becker, M.L. Primary human chondrocyte extracellular matrix formation and phenotype maintenance using rgd-derivatized pegdm hydrogels possessing a continuous young's modulus gradient. *Acta Biomater.* **2013**, *9*, 6095–6104. [CrossRef] [PubMed]

84. Du, Y.; Hancock, M.J.; He, J.; Villa-Uribe, J.; Wang, B.; Cropek, D.M.; Khademhosseini, A. Convection driven generation of long-range material gradients. *Biomaterials* **2010**, *31*, 2686. [CrossRef] [PubMed]

85. Tripathi, A.; Kathuria, N.; Kumar, A. Elastic and macroporous agarose–gelatin cryogels with isotropic and anisotropic porosity for tissue engineering. *J. Biomed. Mater. Res. Part A* **2009**, *90A*, 680–694. [CrossRef] [PubMed]

86. Laco, F.; Grant, M.H.; Black, R.A. Collagen–nanofiber hydrogel composites promote contact guidance of human lymphatic microvascular endothelial cells and directed capillary tube formation. *J. Biomed. Mater. Res. A* **2013**, *101A*, 1787–1799. [CrossRef] [PubMed]

87. Vichare, S.; Sen, S.; Inamdar, M.M. Cellular mechanoadaptation to substrate mechanical properties: Contributions of substrate stiffness and thickness to cell stiffness measurements using afm. *Soft Matter* **2014**, *10*, 1174–1181. [CrossRef] [PubMed]

88. Kühn, P.T.; Zhou, Q.; van der Boon, T.A.B.; Schaap-Oziemlak, A.M.; van Kooten, T.G.; van Rijn, P. Double linear gradient biointerfaces for determining two-parameter dependent stem cell behavior. *ChemNanoMat* **2016**, *2*, 407–413. [CrossRef]

89. Lo, C.T.; Throckmorton, D.J.; Singh, A.K.; Herr, A.E. Photopolymerized diffusion-defined polyacrylamide gradient gels for on-chip protein sizing. *Lab Chip* **2008**, *8*, 1273–1279. [CrossRef] [PubMed]

90. He, J.; Du, Y.; Guo, Y.; Hancock, M.J.; Wang, B.; Shin, H.; Wu, J.; Li, D.; Khademhosseini, A. Microfluidic synthesis of composite cross-gradient materials for investigating cell–biomaterial interactions. *Biotechnol. Bioeng.* **2011**, *108*, 175–185. [CrossRef] [PubMed]

91. Kilian, K.A.; Bugarija, B.; Lahn, B.T.; Mrksich, M. Geometric cues for directing the differentiation of mesenchymal stem cells. *Proc. Natl. Acad. Sci. USA* **2010**, *107*, 4872–4877. [CrossRef] [PubMed]

92. Pedron, S.; Peinado, C.; Bosch, P.; Benton, J.A.; Anseth, K.S. Microfluidic approaches for the fabrication of gradient crosslinked networks based on poly(ethylene glycol) and hyperbranched polymers for manipulation of cell interactions. *J. Biomed. Mater. Res. A* **2011**, *96A*, 196–203. [CrossRef] [PubMed]

93. Lin, N.J.; Bailey, L.O.; Becker, M.L.; Washburn, N.R.; Henderson, L.A. Macrophage response to methacrylate conversion using a gradient approach. *Acta Biomater.* **2007**, *3*, 163–173. [CrossRef] [PubMed]

94. He, J.; Du, Y.; Villa-Uribe, J.L.; Hwang, C.; Li, D.; Khademhosseini, A. Rapid generation of biologically relevant hydrogels containing long-range chemical gradients. *Adv. Funct. Mater.* **2010**, *20*, 131–137. [CrossRef] [PubMed]

95. Burdick, J.A.; Khademhosseini, A.; Langer, R. Fabrication of gradient hydrogels using a microfluidics/photopolymerization process. *Langmuir* **2004**, *20*, 5153–5156. [CrossRef] [PubMed]

96. Zaari, N.; Rajagopalan, P.; Kim, S.K.; Engler, A.J.; Wong, J.Y. Photopolymerization in microfluidic gradient generators: Microscale control of substrate compliance to manipulate cell response. *Adv. Mater.* **2004**, *16*, 2133–2137. [CrossRef]

97. Nemir, S.; Hayenga, H.N.; West, J.L. Pegda hydrogels with patterned elasticity: Novel tools for the study of cell response to substrate rigidity. *Biotechnol. Bioeng.* **2010**, *105*, 636–644. [CrossRef] [PubMed]

98. Vickerman, V.; Blundo, J.; Chung, S.; Kamm, R. Design, fabrication and implementation of a novel multi parameter control microfluidic platform for three-dimensional cell culture and real-time imaging. *Lab Chip* **2008**, *8*, 1468–1477. [CrossRef] [PubMed]

99. Lo, C.M.; Wang, H.B.; Dembo, M.; Wang, Y.L. Cell movement is guided by the rigidity of the substrate. *Biophys. J.* **2000**, *79*, 144–152. [CrossRef]

100. Guarnieri, D.; Borzacchiello, A.; De Capua, A.; Ruvo, M.; Netti, P.A. Engineering of covalently immobilized gradients of rgd peptides on hydrogel scaffolds: Effect on cell behaviour. *Macromol. Symp.* **2008**, *266*, 36–40. [CrossRef]

101. Parekh, S.H.; Chatterjee, K.; Lin-Gibson, S.; Moore, N.M.; Cicerone, M.T.; Young, M.F.; Simon, C.G., Jr. Modulus-driven differentiation of marrow stromal cells in 3D scaffolds that is independent of myosin-based cytoskeletal tension. *Biomaterials* **2011**, *32*, 2256–2264. [CrossRef] [PubMed]

102. Smith Callahan, L.A.; Childers, E.P.; Bernard, S.L.; Weiner, S.D.; Becker, M.L. Maximizing phenotype constraint and extracellular matrix production in primary human chondrocytes using arginine-glycine-aspartate concentration gradient hydrogels. *Acta Biomater.* **2013**, *9*, 7420–7428. [CrossRef] [PubMed]

103. Smith Callahan, L.A.; Policastro, G.M.; Bernard, S.L.; Childers, E.P.; Boettcher, R.; Becker, M.L. Influence of discrete and continuous culture conditions on human mesenchymal stem cell lineage choice in RGD concentration gradient hydrogels. *Biomacromolecules* **2013**, *14*, 3047–3054. [CrossRef] [PubMed]

104. Kloxin, A.M.; Benton, J.A.; Anseth, K.S. *In situ* elasticity modulation with dynamic substrates to direct cell phenotype. *Biomaterials* **2010**, *31*, 1–8. [CrossRef] [PubMed]

105. Marklein, R.A.; Burdick, J.A. Spatially controlled hydrogel mechanics to modulate stem cell interactions. *Soft Matter* **2009**, *6*, 136–143. [CrossRef]

106. Cassereau, L.; Miroshnikova, Y.A.; Ou, G.; Lakins, J.; Weaver, V.M. A 3D tension bioreactor platform to study the interplay between ecm stiffness and tumor phenotype. *J. Biotechnol.* **2015**, *193*, 66–69. [CrossRef] [PubMed]

107. Kapur, T.A.; Shoichet, M.S. Immobilized concentration gradients of nerve growth factor guide neurite outgrowth. *J. Biomed. Mater. Res. A* **2004**, *68*, 235–243. [CrossRef] [PubMed]

108. Yufei, M.; Yuan, J.; Guoyou, H.; Kai, L.; Xiaohui, Z.; Feng, X. Bioprinting 3D cell-laden hydrogel microarray for screening human periodontal ligament stem cell response to extracellular matrix. *Biofabrication* **2015**, *7*, 044105. [CrossRef]

109. DeLong, S.A.; Moon, J.J.; West, J.L. Covalently immobilized gradients of BFGF on hydrogel scaffolds for directed cell migration. *Biomaterials* **2005**, *26*, 3227–3234. [CrossRef] [PubMed]

110. Dodla, M.C.; Bellamkonda, R.V. Anisotropic scaffolds facilitate enhanced neurite extension *in vitro*. *J. Biomed. Mater. Res. A* **2006**, *78A*, 213–221. [CrossRef]

111. Guarnieri, D.; De Capua, A.; Ventre, M.; Borzacchiello, A.; Pedone, C.; Marasco, D.; Ruvo, M.; Netti, P.A. Covalently immobilized RGD gradient on peg hydrogel scaffold influences cell migration parameters. *Acta Biomater.* **2010**, *6*, 2532–2539. [CrossRef] [PubMed]

112. Sarvestani, A.S.; Jabbari, E. Analysis of cell locomotion on ligand gradient substrates. *Biotechnol. Bioeng.* **2009**, *103*, 424–429. [CrossRef] [PubMed]

113. Wu, X.; Newbold, M.A.; Haynes, C.L. Recapitulation of *in vivo*-like neutrophil transendothelial migration using a microfluidic platform. *Analyst* **2015**, *140*, 5055–5064. [CrossRef] [PubMed]

114. Wang, L.; Li, Y.; Chen, B.; Liu, S.; Li, M.; Zheng, L.; Wang, P.; Lu, T.J.; Xu, F. Patterning cellular alignment through stretching hydrogels with programmable strain gradients. *ACS Appl. Mater. Interfaces* **2015**, *7*, 15088–15097. [CrossRef] [PubMed]

115. Uzel, S.G.M.; Amadi, O.C.; Pearl, T.M.; Lee, R.T.; So, P.T.C.; Kamm, R.D. Microfluidics: Simultaneous or sequential orthogonal gradient formation in a 3D cell culture microfluidic platform. *Small* **2016**, *12*, 688–688. [CrossRef]

116. Smith Callahan, L. The concentration game: Differential effects of bioactive signaling in 2D and 3D culture. *Neural Regen. Res.* **2016**, *11*, 66–68. [CrossRef] [PubMed]

117. Wong, A.P.; Perez-Castillejos, R.; Christopher Love, J.; Whitesides, G.M. Partitioning microfluidic channels with hydrogel to construct tunable 3-D cellular microenvironments. *Biomaterials* **2008**, *29*, 1853–1861. [CrossRef] [PubMed]
118. Leipzig, N.D.; Shoichet, M.S. The effect of substrate stiffness on adult neural stem cell behavior. *Biomaterials* **2009**, *30*, 6867–6878. [CrossRef] [PubMed]
119. Norman, L.; Aranda-Espinoza, H. Cortical neuron outgrowth is insensitive to substrate stiffness. *Cell. Mol. Bioeng.* **2010**, *3*, 398–414. [CrossRef]
120. Ali, S.; Wall, I.B.; Mason, C.; Pelling, A.E.; Veraitch, F.S. The effect of Young's modulus on the neuronal differentiation of mouse embryonic stem cells. *Acta Biomater.* **2015**, *25*, 253–267. [CrossRef] [PubMed]
121. Hong, B.; Xue, P.; Wu, Y.; Bao, J.; Chuah, Y.J.; Kang, Y. A concentration gradient generator on a paper-based microfluidic chip coupled with cell culture microarray for high-throughput drug screening. *Biomed. Microdevices* **2016**, *18*, 1–8. [CrossRef] [PubMed]

*gels*

MDPI

*Review*

# Hydrogels as Extracellular Matrix Analogs

Eva C. González-Díaz and Shyni Varghese *

Department of Bioengineering, University of California San Diego, 9500 Gilman Drive, La Jolla, CA 92093, USA; ecgonzal@eng.ucsd.edu
* Correspondence: svarghese@ucsd.edu; Tel.: +1-858-822-7920

Academic Editor: Esmaiel Jabbari
Received: 13 May 2016; Accepted: 25 July 2016; Published: 3 August 2016

**Abstract:** The extracellular matrix (ECM) is the non-cellular component of tissue that provides physical scaffolding to cells. Emerging studies have shown that beyond structural support, the ECM provides tissue-specific biochemical and biophysical cues that are required for tissue morphogenesis and homeostasis. Hydrogel-based platforms have played a key role in advancing our knowledge of the role of ECM in regulating various cellular functions. Synthetic hydrogels allow for tunable biofunctionality, as their material properties can be tailored to mimic those of native tissues. This review discusses current advances in the design of hydrogels with defined physical and chemical properties. We also highlight research findings that demonstrate the impact of matrix properties on directing stem cell fate, such as self-renewal and differentiation. Recent and future efforts towards understanding cell-material interactions will not only advance our basic understanding, but will also help design tissue-specific matrices and delivery systems to transplant stem cells and control their response in vivo.

**Keywords:** hydrogels; extracellular matrix; bioactive materials

## 1. Introduction

Reciprocal interactions of cells with their microenvironment are fundamental to multiple cellular processes that are necessary for tissue development, homeostasis, and regeneration [1]. The key players of the microenvironment are the extracellular matrix, cytokines, and growth factors, as well as neighboring cells. The extracellular matrix is a dynamic ensemble of proteins and proteoglycans, which surround cells, provide anchoring sites, and regulate growth factor signaling [2–5]. While the effect of soluble molecules and growth factors of the microenvironment on cell fate and function is very well understood, our knowledge of the impact of the physicochemical properties of the extracellular matrix is only just beginning to emerge. Emerging evidence has established that the ECM is not just a passive structural support, as previously thought, but is rather an active modulator of various cellular behaviors contributing to tissue morphogenesis and regeneration [1]. The physical and chemical properties of the ECM are tissue-specific and, when negatively perturbed, could contribute to disease progression [6,7]. Biomaterials have played a key role towards our current understanding of the contribution of matrix properties on the cell response. Among different forms of biomaterials, hydrogels have been widely used as three-dimensional (3D) structural supports to culture cells as they provide a highly-hydrated, cytocompatible environment and facilitate nutrient and waste transport [8].

Hydrogels are three-dimensional networks of hydrophilic polymer chains that can imbibe large quantities of biological fluid. Thus, hydrogels are very similar in structure to the ECM of mammalian tissues, which essentially consists of hydrated proteins and polysaccharide networks. Crosslinking of polymeric chains to form hydrogels can be achieved through covalent or non-covalent interactions [8]. Hydrogels have been used to support both monolayer (2D) and 3D cell culture. While most of the initial studies focused on the ability of hydrogels to provide structural support to cells, recent efforts have focused on recapitulating various physicochemical cues of the native ECM. Hydrogels can be created from biologically-derived, naturally occurring, or synthetic precursors [8]. Hydrogels made from biological precursors, such as collagen, possess biochemical cues relevant to various cellular functions such as attachment, growth, and migration. On the other hand, hydrogels made from synthetic precursors are often biologically inert and lack bioactive moieties necessary for supporting cell adhesion. Modification of synthetic hydrogels with bioactive molecules like proteins or peptides is needed to elicit cell adhesion [9,10]. One of the advantages of hydrogels made from synthetic precursors over their biologically-derived counterparts is that their physical properties (e.g., mechanical and degradation properties) can be easily controlled and tuned. The fact that the physicochemical and biological properties of synthetic hydrogels can be easily altered in a reproducible manner makes them ideal candidates to study the effect of ECM properties on cellular functions in a systematic manner.

This review discusses the design of hydrogels with defined physical, chemical, and tunable spatiotemporal features. The review also summarizes how hydrogels with defined surface and bulk properties can be used to regulate various cell functions, including self-renewal and differentiation of stem cells (Figure 1) as well as tissue formation.

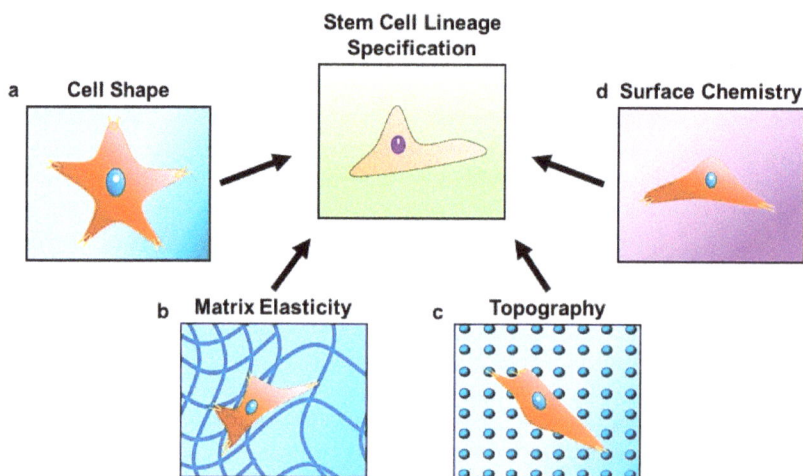

**Figure 1.** Stem cell lineage specification is regulated by changes in (**a**) cell shape dictated by the surrounding matrix; (**b**) matrix elasticity; (**c**) topography; and (**d**) chemical composition at the cell-material interface.

## 2. Designing Hydrogels with Defined Physicochemical Properties

Hydrogel properties can be largely divided into surface and bulk properties. These properties can act in concert or individually to regulate various cell functions. In the following sections, we highlight current strategies in the design of hydrogel matrices with tunable surface and bulk properties and their influence on cell behavior.

### 2.1. Design and Synthesis of Cell-Adhesive Hydrogels through Peptide and Protein Immobilization

Hydrogels made from synthetic hydrophilic polymers generally do not support cell adhesion as they are highly resistant to protein adsorption. In fact, the antifouling properties of hydrophilic polymers have been extensively used to improve the longevity of implants, where the modified surfaces prevent protein adsorption and, thereby, failure of the implants. For example, hydrophilic poly(ethylene glycol) (PEG) coatings have been applied to the surface of poly-urethane arterial shunts to reduce clotting resulting from protein adsorption [11]. While such antifouling surfaces/interfaces are advantageous for improving the function of certain implants, this antifouling property limits the application of hydrogels as cell culture substrates, as the minimum requirement for the survival of anchorage-dependent cells is adhesion to the underlying matrix.

A commonly used approach to make hydrogels cell-adhesive is the incorporation of peptides or proteins into the hydrogel network [10,12–16]. The biofunctionalization of hydrogels can be achieved through bioconjugation, such as coupling between –NHS and amine groups [15], Michael-type addition [13], thiol-acrylate reaction [12], or copolymerization [16]. In addition to being inexpensive and easy to synthesize, peptide-immobilized biomaterials are more amenable to sterilization, unlike proteins that could undergo denaturation at non-physiological conditions. The biological activity of immobilized peptides and proteins relies upon their accessibility, suggesting that the tethered groups must be flexible and experience minimal steric hindrance [13]. The most widely studied cell-adhesive peptide is arginine-glycine-aspartate (or RGD), a tri-amino acid sequence. RGD is the key integrin-binding domain present among different ECM proteins [9].

While RGD sequences can assist attachment of cells to biologically inert hydrogels by engaging cell surface integrins, immobilization of RGDs is often not sufficient to regulate biological and signaling events relevant to maintaining self-renewal or directing differentiation of stem cells. This is not surprising, given that the native ECM presents different bioactive units with varying conformation and densities to modulate cellular functions. A number of studies have incorporated peptide units and different ECM components into hydrogels to achieve targeted cellular functions. For instance, Musah et al., endowed poly(acylamide) (PAm) hydrogels with vitronetin-derived peptide units (GKKQRFRHRNRKG) to interact with cell surface glycans [10]. These modified PAm hydrogels displaying glycosaminoglycan binding peptides were found to support self-renewal of human embryonic stem cells (hESCs) in 2D culture [14].

Similar to proteins, polysaccharides of the native ECM also play an important role in mediating cell-matrix interactions. Native ECM proteoglycans are known to regulate growth factor signaling through sequestration and release of growth factors upon cellular demand [5]. Hence, hydrogels have also been designed to regulate growth factor signaling relevant to various cellular functions [17–19]. This can be achieved by either functionalizing synthetic hydrogels with proteoglycan moieties [20] or synthetic molecules that mimic proteoglycan functions [17,21,22]. For instance, heparin-functionalized PEG hydrogels have been used to direct osteogenic differentiation of human mesenchymal stem cells (hMSCs) in 3D culture [23]. Heparin is known to sequester growth factors and proteins, such as BMP-2 and fibronectin. Similarly, studies have used hydrogel mineralization to sequester growth factors, such as BMP-2 [24]. The incorporation of chondroitin sulfate (CS) and hyaluronic acid moieties into PEG hydrogels has also been used to impart bioactivity [20,25,26]. A study by Varghese et al. demonstrated that hMSCs encapsulated within CS/PEG hydrogels promoted the formation of cell aggregates and enhanced chondrogenic differentiation and deposition of cartilage-specific ECM [27]. Hydrogels containing hyaluronic acid moieties have been shown to promote chondrogenic differentiation of stem cells, such as MSCs, in 3D cultures [28]. In addition to being a key component of cartilage ECM, hyaluronic acid molecules may interact with encapsulated MSCs through CD44 receptors [29]. Aside from their use as scaffolds for cartilage tissue engineering, hyaluronic acid hydrogels have been explored for a variety of other biomedical applications [30].

While active conjugation of peptides and ECM components can be used to impart bioactivity, non-specific adsorption of proteins can also make hydrogels adhesive to assist cell attachment. A number of parameters, such as surface roughness, chemistry, and hydrophobicity (or hydrophilicity), influence the ability of hydrogels to adsorb proteins [11].

*2.2. Tuning the Cell-Matrix Interface through Functional Groups and Hydrophobicity*

Hydrogels displaying certain functional groups and hydrophobicity can support cell culture in the absence of active immobilization of peptides or proteins, often through non-specific protein adsorption [31]. In general, hydrophobic surfaces have a higher tendency for protein adsorption than do hydrophilic surfaces [11,32]. The adsorbed proteins at the cell-material interface provide cell adhesive domains that support cell attachment and growth. Surface chemistry and hydrophobicity not only influence protein adsorption, but also modulate protein conformation, which in turn regulates integrin binding and cell function [31,33]. In addition to supporting survival and growth, these parameters also play a key role in cell migration and differentiation of stem cells [33,34].

By incorporating small-molecules into PEG hydrogel matrices, Benoit et al. examined the effect of matrix functional groups on hMSC differentiation (Figure 2) in 3D culture [35]. Hydrogels containing phosphate functional groups induced osteogenic differentiation, whereas those containing *t*-butyl groups promoted adipogenic differentiation. Additionally, gels that were functionalized with methacrylic acid stimulated the upregulation of cartilage-specific markers, ultimately leading to chondrogenic differentiation. The molecular mechanism by which chemical functional groups induce differentiation of stem cells into a particular phenotype remains unknown.

Hydrogel functional groups have also been used to generate synthetic matrices with bone-specific biochemical cues (i.e., mineral components) [36,37]. In a recent study, we used hydrogels with carboxyl functional groups to generate matrices bearing calcium phosphate (CaP) minerals [38]. These biomineralized hydrogels were found to direct osteogenic commitment of stem cells, such as hMSCs, hESCs, and human induced pluripotent stem cells (hiPSCs), in 2D and 3D cultures, in the absence of any other osteogenic molecules [39–41]. The dynamic dissolution (into $Ca^{2+}$ and $PO_4^{3-}$ ions) and re-precipitation of matrix-bound CaP minerals has been touted to play a key role in the osteoinductivity of these mineralized matrices. This dynamic dissolution and re-precipitation of CaP minerals not only modulates $Ca^{2+}$ and $PO_4^{3-}$ signaling to influence osteogenic differentiation [42–44], but can also sequester and release osteoinductive growth factors such as bone morphogenic proteins (BMPs) [24]. In addition, the CaP minerals of the matrix could contribute to osteogenic commitment of stem cells through $PO_4^{3-}$-ATP-Adenosine-A2b receptor axis signaling [42] while inhibiting their differentiation into adipogenic lineage [43].

A study by Phillips et al. sought to understand the effect of functional groups on hMSC differentiation by using self-assembled monolayer (SAM) surfaces [34]. Four functional groups: $CH_3$, OH, COOH, and $NH_2$ were used to represent hydrophobic, hydrophilic, negatively charged, and positively charged interfacial properties, respectively. In this study, surfaces functionalized with –OH and –$NH_2$ demonstrated a strong upregulation of osteogenic markers along with a downregulation of adipogenic markers, while showing no significant effect on chondrogenic differentiation. Valamehr et al. used SAM surfaces to examine the effect of substrate hydrophobicity on differentiation of embryonic stem cell-derived embryoid bodies [45]. In another study, hydrogels with low wettability (hydrophobic surfaces) were shown to support clonal growth of hESCs and hiPSCs in 2D culture, through non-specific adsorption of vitronectin [46]. The vitronectin adsorbed onto the surfaces engaged with the cells through $\alpha_v\beta_3$ and $\alpha_v\beta_5$ integrins and assisted their growth while maintaining pluripotency. A study by Chang et al. incorporated styrene sulfonate functional groups, a potential synthetic analog of heparin, to regulate bFGF signaling and generate hydrogels that could support human pluripotent stem cell growth in monolayer culture and maintain pluripotency ex vivo [21].

While the aforementioned studies demonstrate the influence of matrix interfacial properties on determining various cellular outcomes, it is often difficult to decouple the effect of functional groups and hydrophobicity. A recent study by Ayala et al. addressed this issue by using acryloyl amino acid (AA) monomers with varying side chain lengths (through the number of –$CH_2$ groups that separate the vinyl group and terminal –COOH group) [33]. Acrylamide hydrogels functionalized with AA units of varying chain length showed different levels of hydrophobicity without altering the stiffness or the hydrogel functional group. The results from this study showed cells adhered to hydrogels exhibiting an optimal hydrophobicity which grew to confluence, where the adhesion of cells to the underlying matrix was mediated through nonspecific protein adsorption. Based on the shape and alignment of the adhered cells, the cultured hMSCs underwent either osteogenic or adipogenic differentiation.

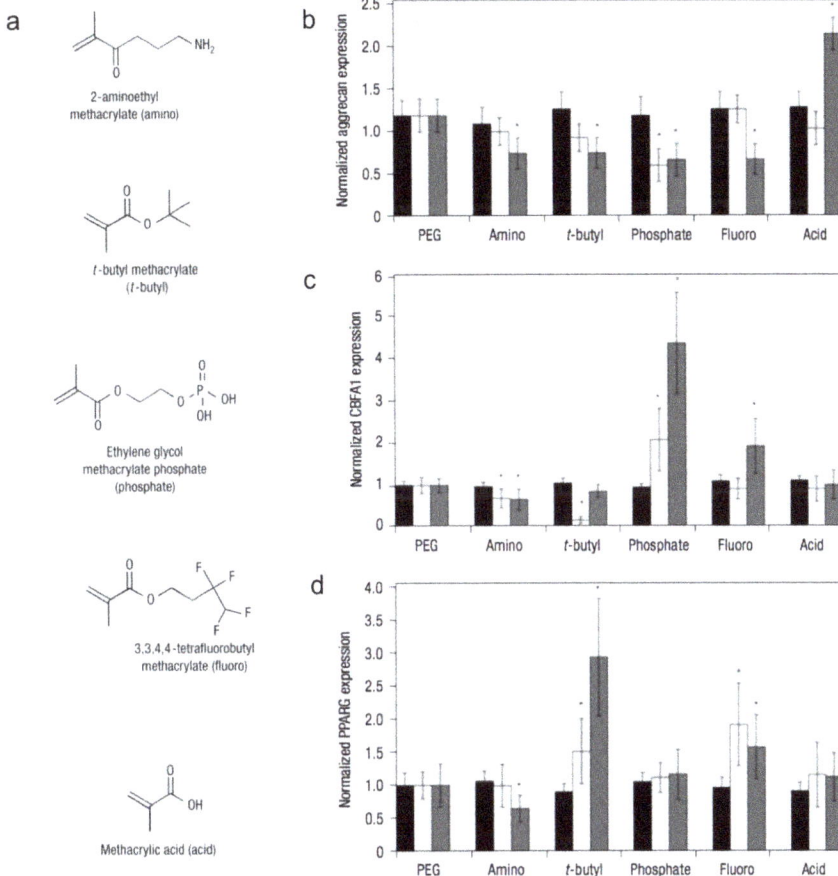

**Figure 2.** Small-molecule incorporation alters human mesenchymal stem cell (hMSC) gene expression on poly(ethylene glycol) (PEG) hydrogels. (**a**) Chemical structures of functional moieties incorporated. Gene expression of hMSCs (as measured by in situ hybridization) quantitatively analyzed for aggrecan (**b**); CBFA1 (**c**); and PPARG (**d**) at days 0 (black bars), 4 (white bars) and 10 (grey) of culture on unmodified PEG and 50 mM of amino, *t*-butyl, phosphate, fluoro, and acid. Values are reported as the fluorescent intensity average of six samples per composition per time point, relative to β-actin expression, and normalized to expression by cells cultured on PEG surfaces. Error bars represent one standard deviation. An asterisk (*) denotes statistical significance compared with PEG ($p < 0.05$). Adapted with permission from [35]. Copyright 2008 Nature.

The hydrophobicity-mediated "adhesivity" of hydrogels has been used to develop "smart surfaces" for cell culture [47,48]. Hydrogels displaying smart surfaces are generally fabricated from polymers exhibiting lower critical solution temperature (LCST). Thus far, poly(N-isopropylacrylamide) (pNIPAm) is the most widely-used, temperature-responsive polymer for cell culture. Thermoresponsive hydrogels oscillate between hydrophilic and hydrophobic surfaces around the LCST temperature [49]. At 37 °C, the hydrogel surface is hydrophobic, enabling nonspecific protein adsorption and making the surface cell adhesive. At temperatures below 37 °C, the hydrogel surface becomes hydrophilic and releases the monolayer of cells as a sheet. Such engineered cell sheets have been used to treat a myriad of disorders. For example, engineered myocardial cells sheets have been developed to treat patients suffering from severe heart failure [50]. Such stimuli-responsive hydrogels have also been used for minimally-invasive cell delivery [51], expansion of pluripotent stem cells [52], and multi-functional scaffolds for cell culture [53].

### 2.3. Design of Hydrogels with Topographical Cues

In the human body, the native environment that cells experience is far from flat. The organization of the extracellular matrix gives rise to complex geometrical features, which play a significant role in various cellular functions. A number of studies have employed micropatterned, cell-adhesive geometrical features of various sizes and shapes to examine the effect of cell shape on growth, polarization, migration, and differentiation of stem cells [54–56]. Throughout this section, we discuss the most commonly used techniques to generate micropatterned matrices and highlight key studies that have served to expand our understanding of the cell response to topographical cues in 2D. One common micropatterning technique is photolithography, which consists of polymerizing a material by exposing the hydrogel precursor solution to ultraviolet (UV) light through a photomask displaying the desired pattern [57]. A similar approach has been applied to generate hydrogel patterns within microfluidic chips [58,59]. Soft lithographic techniques, such as microcontact printing, involve the use of a master stamp, often made from polydimethylsiloxane (PDMS), that can transfer adhesive proteins or other molecules (referred to as the ink) onto a substrate, to generate patterned features [60,61]. Another approach is micromolding, where the hydrogel precursor solution is placed over a PDMS stamp containing the negative pattern of the desired geometry and allowed to polymerize (usually by exposure to UV light) [62]. The stamp is subsequently removed and the hydrogel is inverted to reveal the patterned surface. Another technique used to generate matrices with topographical features is electrospinning [63]. In this technique, a charged stream of polymer solution is placed within a syringe and exposed to an electric field. The voltage is increased until the electric force generated overcomes the surface tension at the tip of the needle, resulting in the ejection of a jet of polymer that can then be collected on a rotating or stationary collector in the desired orientation. Studies have also used differential swelling of hydrogels as a tool to create surface wrinkles, which can be used to generate different topographical features [55,64].

Findings from studies using micropatterned matrices have shed light into the remarkable manner with which the cytoskeletal architecture of the cell adapts to the shape provided by the substrate surface, subsequently influencing its migration, growth, and differentiation. Specifically, the size and shape of the patterned domains governs the cell volume and spreading, the organization of cytoskeletal networks and, subsequently, intracellular signaling. Results from these efforts have shown that the commitment of MSCs to either an osteogenic or adipogenic lineage can be regulated by the cell shape [65,66]. For instance, cells cultured on the surface of large islands demonstrated increased adhesion and spreading and eventually underwent osteogenesis. In contrast, cells on smaller islands underwent adipogenesis after adopting a round and unspread morphology [65]. The effect of matrix topographical cues on stem cell commitment was further demonstrated by subsequent studies [66] in which geometrical constraints leading to increased actomyosin contractility directed osteogenic differentiation, while those of low contractility led to adipogenic differentiation. Most of these initial seminal studies utilized PDMS, a crosslinked hydrophobic polymer, as a substrate.

In recent years, these efforts have been extended towards the creation of hydrogels with topographical features. For instance, Lee et al. studied the effect of geometric confinement on MSC spreading and lineage specification by patterning ECM proteins, such as fibronectin, laminin, and type I collagen, over the surface of hydrazine-treated polyacrylamide hydrogels [67]. Cells that were more spatially constrained adopted a round morphology and ultimately underwent adipogenic differentiation. In contrast, cells that were allowed to spread freely over the hydrogel surface showed upregulation of neurogenic markers. A similar study used micropatterned polyacrylamide hydrogels to demonstrate the effect of 2D geometric cues on osteogenic differentiation of MSCs [68]. In this study, osteogenesis was enhanced when cells where cultured on geometric shapes that generated an increase in cytoskeletal tension, as was observed for cells growing on elongated shapes (Figure 3).

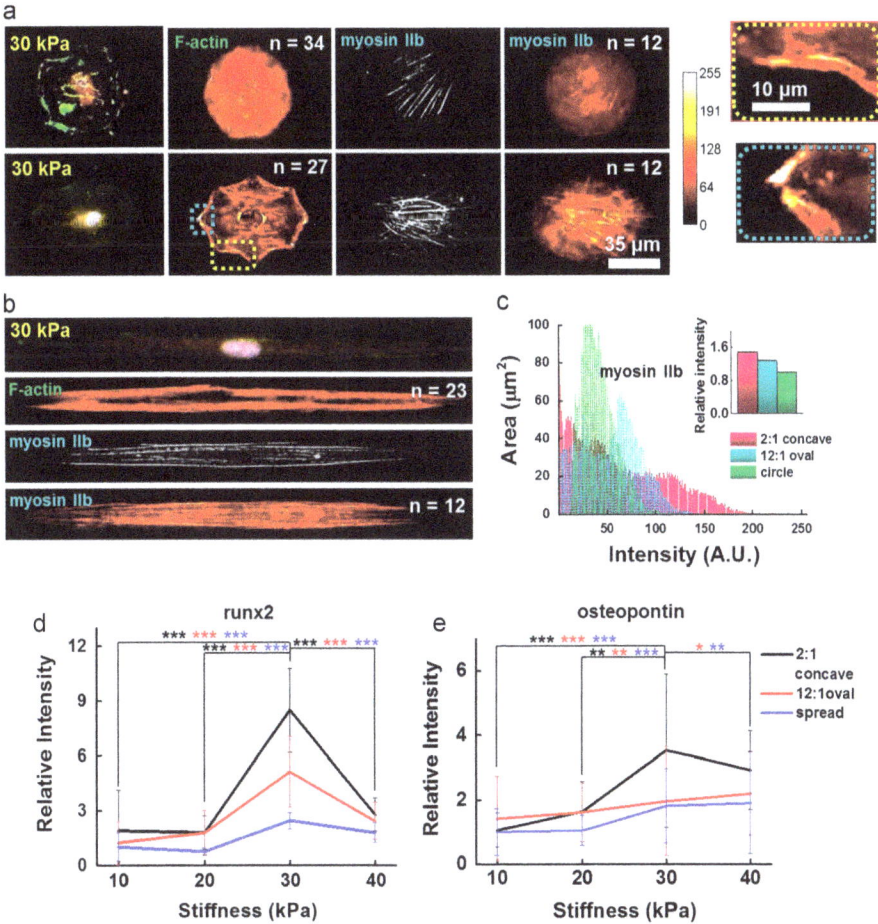

**Figure 3.** Influence of cell shape on the MSC cytoskeleton. (**a**) and (**b**) show immunofluorescence images and immunofluorescence heatmaps (left to right: F-actin with nuclei, heatmap of F-actin IIb, and heatmap of myosin IIb) for cells cultured in circular, concave, and elongated shape for 10 days; (**c**) heat map intensity comparison for cells stained for myosin IIb. Inset represents myosin IIb intensity normalized to that of circular geometry. Additionally, enhanced osteogenesis marker expression was observed in mesenchymal stem cells patterned in contractile geometries; (**d**) Relative runx2 marker intensity of cells captured on concave or oval shapes or spread on the fibronectin matrix protein, differentiating to osteogenic lineages (*** $p < 0.0005$, *t*-test compared to concave cells on 30 kPa). Runx2 nuclear fluorescence was normalized to cytoplasmic fluorescence. The relative intensity of the fluorescence was determined by comparing each intensity value to the average intensity of spread cells on 10 kPa; (**e**) Relative osteogenic marker intensity (osteopontin) of cells captured on concave or oval shapes or spread on the fibronectin matrix protein (* $p < 0.05$, ** $p < 0.005$, *** $p < 0.0005$, *t*-test compared to concave cells on 30 kPa). The relative intensity of the fluorescence was determined by comparing each intensity value to the average intensity of spread cells on 10 kPa. Adapted with permission from [68]. Copyright 2014 Elsevier.

Surface grates, commonly consisting of parallel lines of defined width and depth, have been employed to enhance cell adhesion and guide cell polarization [69,70]. Specifically, this technique has proven successful in directing neurite extension of PC12s, a neuronal progenitor cell line, as these extensions form parallel to the axis of the grates [71]. Additionally, seeding hMSCs over nanogrates that were 350 nm wide resulted in an upregulation of microtubule-associated protein 2 (MAP2), a key marker in neuronal differentiation [72]. Using a similar concept, nanopitted surfaces have been used to study hMSC differentiation. Interestingly, identical pit dimensions can be used for entirely different purposes, ranging from stem cell maintenance [73] to osteogenesis [74], by simply varying their spatial arrangement.

## 3. Design of Hydrogel Bulk Properties to Probe and Direct Cell Function

### 3.1. Tuning Matrix Stiffness to Guide Cell Behavior

The human body is comprised of tissues with vastly different mechanical properties, ranging from soft tissue, such as that found in the brain, to the stiff tissue that constitutes bone. This has led to activities examining the role of matrix mechanical properties on cell and tissue functions. Hydrogels have been extensively used to study the effect of matrix mechanical properties, such as Young's modulus (commonly termed as stiffness) on cell function both in vitro and in vivo. Hydrogel matrix stiffness can be varied by controlling the network crosslink density. The network crosslink density can be increased by increasing the concentrations of the crosslinker and/or the monomer or oligomer concentration [75]. Other approaches to improve the mechanical properties of hydrogels include the incorporation of hydrophobic domains (to control swelling) [76], nanoclays (which act as physical crosslinks) [77,78], sacrificial chains, or by unzipping ionic crosslinks (such as in the case of double-network hydrogels) [79,80].

In vitro studies using hydrogels have demonstrated that matrix mechanical properties play a crucial role in stem cell phenotypic expression by influencing cell shape and mechanotransduction [81]. Essentially, matrix stiffness has been shown to influence various cell functions, including cell adhesion [82,83], proliferation [84], migration [82,83], and stem cell differentiation [85]. For example, proliferation of neural stem cells in 3D hydrogels increases when the elastic modulus is decreased from ~20,000 Pa to ~180 Pa [84]. Stiffer matrices also allow for stronger cell adhesion and decrease the rate of cell migration [86]. Two dimensional studies have shown that when cultured on hydrogels exhibiting a stiffness gradient, cells undergo directed migration towards higher matrix stiffness [82]. A seminal study performed by Engler et al. demonstrated that matrix elasticity can play a key role in directing stem cell differentiation (Figure 4) in 2D culture [85]. The authors showed that preconditioned hMSCs that were cultured on hydrogels with an elastic modulus (E) ranging from 0.1–1 kPa underwent neurogenesis, while those cultured on stiffer hydrogels of modulus ranging from 8–17 kPa and 25–40 kPa underwent myogenesis and osteogenesis, respectively. Similarly, a number of studies have documented the importance of matrix mechanical properties on maintaining cellular functions [83,87,88].The effect of matrix stiffness is also evaluated in 3D culture [89,90]. Studies by Khetan et al. [90] utilizing degradable hydrogels have demonstrated the role of traction stresses generated by encapsulated hMSCs on their fate commitment. Essentially, hydrogel networks that permitted high traction stresses of hMSCs supported their osteogenic differentiation, whereas those with low traction stresses stimulated adipogenic differentiation. Cells respond to matrix rigidity by exerting traction forces on the surrounding matrix through focal adhesions. These integrin binding sites serve as a line of mechanical communication with the cell cytoskeleton, such that increased resistance to deformation in the matrix is reflected by an increase in cytoskeletal tension. Changes in cytoskeletal tension and actomyosin contractility have been shown to trigger various signaling cascades, such as RhoA signaling, that influence transcriptional regulation of associated genes. For instance, an increase in RhoA signaling has been shown to direct osteogenic commitment of MSCs, while a decrease in RhoA promotes adipogenic differentiation [65].

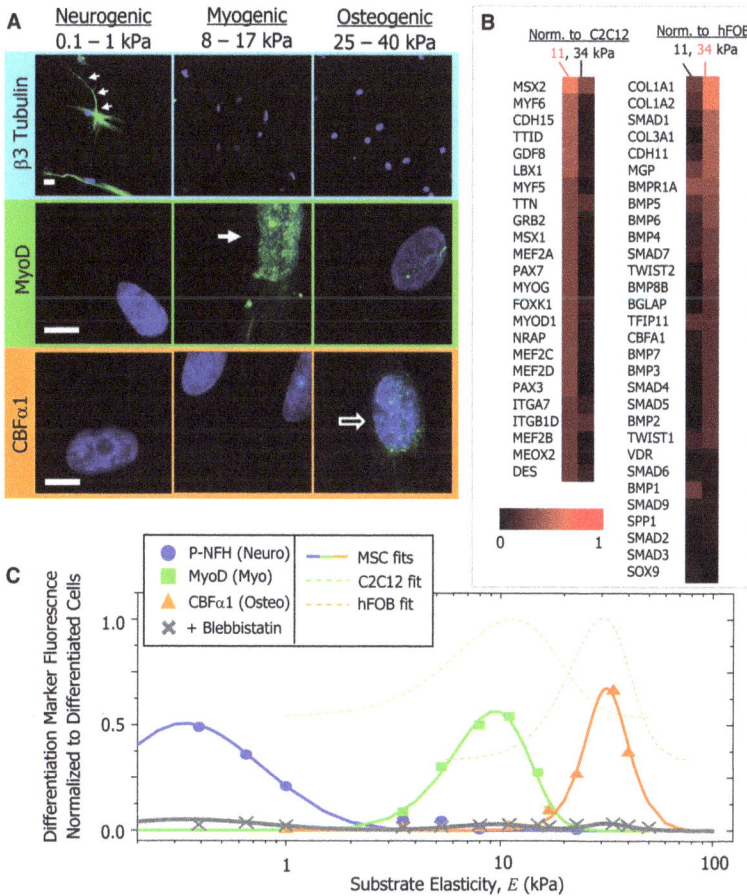

**Figure 4.** Protein and transcript profiles are elasticity dependent under identical media conditions (**A**) The neuronal cytoskeletal marker b3 tubulin is expressed in branches (arrows) of initially naive MSCs (>75%) and only on the soft, neurogenic matrices. The muscle transcription factor MyoD1 is upregulated and nuclear localized (arrow) only in MSCs on myogenic matrices. The osteoblast transcription factor CBFa1 (arrow) is likewise expressed only on stiff, osteogenic gels. Scale bar is 5 mm; (**B**) Microarray profiles of MSCs cultured on 11 or 34 kPa matrices, with expression normalized first to actin and then to expression of committed C2C12 myoblasts and hFOB osteoblasts; (**C**) Fluorescent intensity of differentiation markers versus substrate elasticity reveals maximal lineage specification at the *E* typical of each tissue type. Average intensity is normalized to peak expression of control cells (C2C12 or hFOB). Adapted with permission from [85]. Copyright 2006 Elsevier.

In native tissue, the traction forces that cells exert on the surrounding ECM, along with the mechanical properties of the matrix, dictate the extent to which cells are able to remodel their environment. In turn, the resistance to traction forces decreases over time, thus influencing cell behavior [91]. Recently, Chaudhuri et al. created reversible, 3D alginate hydrogels with stress relaxation properties to understand the effect of non-linear mechanical properties of the ECM on cell functions [92]. Hydrogels with a faster rate of stress relaxation not only improved cell spreading [91,92] and proliferation, but also induced osteogenic differentiation of MSCs (Figure 5) [92]. The mechanical properties of the matrix also have a significant effect in local clustering of RGD ligands, actomyosin contractility, as well as the nuclear translocation of YAP (Yes-associated protein), a key transcriptional regulator involved in stem cell differentiation [93,94].

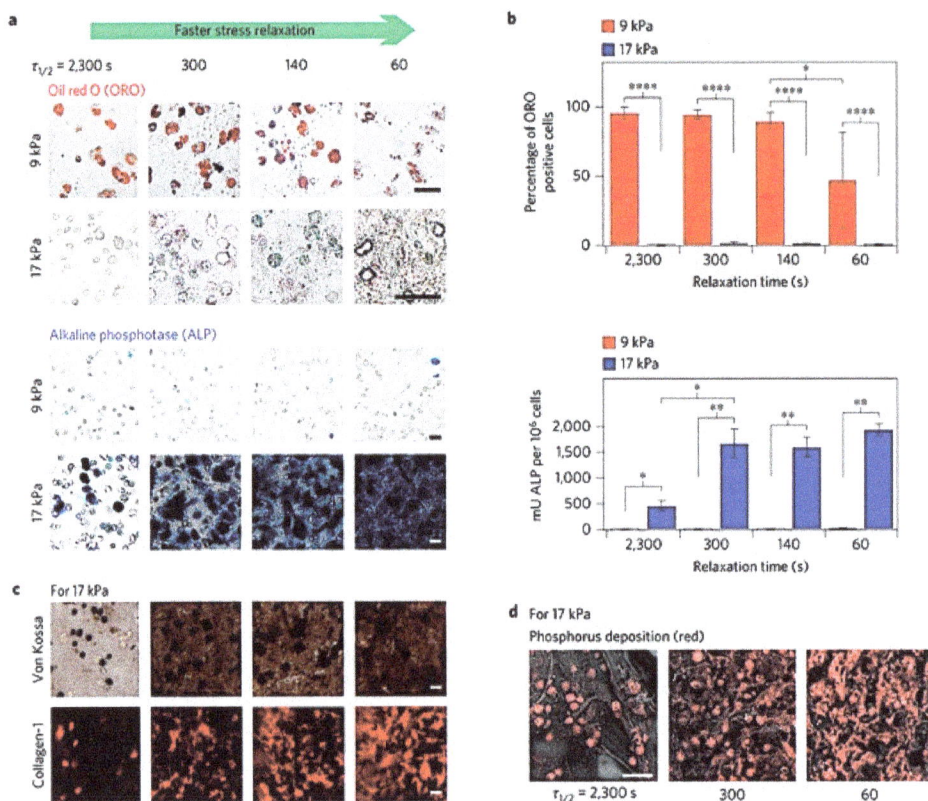

**Figure 5.** MSCs undergo osteogenic differentiation and form an interconnected mineralized collagen-1-rich matrix only in rapidly relaxing gels. (**a**) Oil Red O staining (red), indicating adipogenic differentiation, and alkaline phosphatase staining (blue), indicating early osteogenic differentiation, for MSC cultured in gels of indicated initial modulus and timescale of stress relaxation for seven days. Scale bars are 25 μm; (**b**) Percentage of cells staining positive for Oil Red O, and a quantitative assay for alkaline phosphatase activity. *, **, and **** indicate $p < 0.05$, 0.01, and 0.0001 respectively (Student's *t*-test); (**c**) Von Kossa (mineralization) and collagen-1 stain on cryosections from gels with the indicated conditions after two weeks of culture. Scale bars are 25 μm; (**d**) Scanning electron microscope and energy-dispersive X-ray spectrometry (SEM-EDS) images of sections of gels with the indicated conditions after two weeks of 3D culture of MSCs. Phosphorus elemental maps (P mapped in red) are overlaid on their corresponding backscattered SEM images. Scale bar is 50 μm. Adapted with permission from [92]. Copyright 2015 Nature.

Efforts have also been made to dynamically tune matrix elasticity and recapitulate certain dynamic features of native ECM. To this end, Stowers et al. have created 3D alginate hydrogels (through Ca$^{2+}$ mediated gelation) embedded with light-sensitive liposomes [95]. Encapsulation of either calcium or DTPA (a chelating agent) into these liposomes allowed for the light-triggered and spatially-controlled stiffening or softening of the hydrogel. This design was used to investigate the effect of hydrogel stiffening dynamics on cell morphology using 3T3 fibroblasts. Hydrogels containing encapsulated cells were irradiated for 30, 60, and 120 s, resulting in an increase in stiffness that was proportional to the irradiation time. While fibroblasts that were encapsulated in the unirradiated hydrogel (control group) retained an elongated morphology, cells in the irradiated hydrogels exhibited a round morphology when adapting to the increasing hydrogel stiffness. As a proof of concept, light-triggered stiffening of the hydrogels was also successfully achieved in vivo through transdermal irradiation of the constructs after subcutaneous implantation in mice. In addition, other approaches have been used, such as thiolene polymerization [96] and incorporation of photocleavable moieties [97], to manipulate hydrogel mechanical properties post-encapsulation in culture.

Similarly, a study by Guvendiran and Burdick investigated the effect of dynamic substrate stiffening by using methacrylated hyaluronic acid [98]. Hydrogels were initially crosslinked through Michael-type addition reaction using dithiothreitol (DTT) at various concentrations to tune the initial matrix stiffness (between ~3 and ~100 kPa). The unreacted methacrylate groups were subsequently used to increase the stiffness of the hydrogels through photopolymerization. Within 4 h, both the mean area and the average traction of the encapsulated hMSCs increased in response to an increase in stiffness in the hydrogel network (from ~3 to ~30 kPa). Additionally, the hMSCs differentiated into different phenotypes depending on the amount of time they were cultured in either soft or stiff hydrogels. Adipogenic differentiation was observed with late stiffening of the matrices while osteogenic differentiation was observed in cells cultured in hydrogels with early stiffening.

## 3.2. Designing Pore Architecture to Promote Tissue Formation

Efficient nutrient and waste transport is key to long term cell survival and tissue formation. Hence, matrix porosity plays a significant role in 3D cell culture and engineering of functional tissues from stem cells. Furthermore, matrices with porosity and interconnectivity have also been shown to promote host cell infiltration, homogeneous cell distribution, and integration with the surrounding native tissue. Throughout this section, we highlight recent studies investigating the effect of pore architecture on cell function and tissue formation in 3D.

Porous hydrogels can be generated using a variety of methods, such as solvent casting and particle leaching [99,100], freeze-drying [101], electrospinning [102–104], and gas foaming [105]. Pore architecture must be chosen with a tissue-specific context in mind to improve cell function. For instance, Zeng et al. studied the effect of pore size on chondrocyte growth and function using microcavitary alginate hydrogels [106]. Porcine chondrocytes were encapsulated within matrices of various pore size ranges: 80–120 μm, 150–200 μm, and 250–300 μm. After 21 days, cells that were cultured in hydrogels with pore sizes of 80–120 μm exhibited better growth and maintenance of the chondrocyte phenotype.

The pore architecture (pore size, porosity, and pore interconnectivity) of the matrix must also facilitate cell infiltration and angiogenesis when implanted in vivo. Angiogenesis is particularly important for maintaining cell viability and promoting integration of the engineered tissue with the host tissue. The importance of vascularization becomes increasingly apparent in therapeutic strategies that involve cell transplantation, as poor cell survival often limits the potential benefits of the implant. Oliviero et al. developed VEGF-loaded, porous PEG-co-heparin hydrogels to promote angiogenesis and reported that using a pore size range of 35 to 100 μm and a total porosity of ~50.8% is optimal for promoting neovascularization [107]. Similarly, Dziubla et al. investigated the effect of pore size and porosity of poly(2-hydroxyethylmethacrylate) (PHEMA)-based hydrogels on in vitro tubule formation of human microvascular endothelial cells (HMVECs) [108]. Gels were synthesized with pore sizes ranging from ~5 to ~16 μm and with porosities ranging from ~55% to ~90%. While hydrogels with pore sizes lower than ~8 μm showed minimal cell infiltration and vascularization, those with average pore size above ~9 μm and having ~85% porosity or higher exhibited optimal tubule formation and penetration throughout the structures. These tubules were ~7.5–7.85 μm in diameter and had an average tubule length of ~88–102 μm. In another study, Matsiko et al. investigated the effect of matrix pore size on chondrogenic differentiation of MSCs using collagen-hyaluronic acid hydrogels. Their studies showed that matrices with a mean pore size of 300 μm promoted proliferation and chondrogenic differentiation of MSCs when compared to those with smaller mean pore size (94 and 130 μm) [109]. It is important to note that while vascularization is essential for functional engineering of most tissues, matrix architecture should be designed to avoid angiogenesis when dealing with avascular tissues, such as cartilage.

Another key design consideration for implant success is pore interconnectivity. A study by Bakshi et al. used PHEMA hydrogels with interconnected pores of size ranging from 10 to 20 μm and examined their effect on axonal regeneration in vivo. After soaking in brain-derived neurotrophic factor (BDNF), these constructs not only demonstrated significant angiogenesis, but also served as a "bridge" to promote axonal penetration and regeneration after spinal cord injury [110]. Phadke et al. investigated the effect of pore architecture on osteogenic differentiation of hMSCs [111]. Specifically, CaP-mineralized PEGDA-*co*-A6ACA hydrogels were generated with either a randomly-oriented, "spongy" pore architecture (~50–60 μm pore size) or a directional, columnar pore structure (~100–150 μm pore size). hMSCs that grew on spongy cryogels demonstrated a more spread morphology and showed a higher upregulation of osteogenic markers, such as RUNX2, osteopontin, and osteocalcin in vitro.

Aside from pore architecture, the degradation of the matrix also plays a key role in determining tissue formation [112,113]. In an ideal scenario, a scaffold is expected to degrade at a rate that will accommodate cell-secreted ECM without impeding the production of ECM by the cells. A number of different approaches can be adopted to achieve scaffold degradation. This includes incorporation of functional groups, such as poly(esters), that are labile to hydrolytic degradation [114], peptide sequences that are cleavable by proteases, such as matrix metalloproteinases (MMPs) [115], plasmin [116], and elastase [117], and functional moieties that are labile to cell secreted molecules, such as glutathione or thiol-group containing molecules [118].

## 4. Conclusions

Recent fundamental and technological advancements have significantly improved our understanding of the active participation of hydrated ECM on various cellular functions, ranging from survival to phenotypic commitment. Improvements in the formulation of biomimetic hydrogels that incorporate tissue-specific biochemical and biophysical cues to control stem cell lineage specificity in vitro and in vivo have been truly dramatic in recent years. Beyond expanding our basic understanding of stem cell biology, many of these developments have significantly advanced the field of regenerative medicine and its prospect of moving from the bench to the bedside. However, widespread clinical application of these advancements still relies on our ability to standardize the manufacturing and scale-up processes. Nonetheless, there is no doubt that novel methods at the interface of biomaterial manipulation and stem cell biology will continue to be successfully used to propel the advancement of regenerative medicine and translation of stem cell-based therapeutics.

**Acknowledgments:** The authors acknowledge the financial support from National Institute of Arthritis and Musculoskeletal and Skin Diseases of the National Institutes of Health under Award Number R01 AR063184.

**Author Contributions:** Eva C. González-Díaz gathered the literature material, wrote the review and assembled the figure panels. Shyni Varghese organized the topics, contributed to the writing and discussion of the contents.

**Conflicts of Interest:** The authors declare no conflict of interest.

## References

1. Nelson, C.M.; Bissell, M.J. Of extracellular matrix, scaffolds, and signaling: Tissue architecture regulates development, homeostasis, and cancer. *Annu. Rev. Cell Dev. Biol.* **2006**, *22*, 287–309. [CrossRef] [PubMed]
2. Geiger, B.; Bershadsky, A.; Pankov, R.; Yamada, K.M. Transmembrane extracellular matrix-cytoskeleton crosstalk. *Nat. Rev. Mol. Cell Biol.* **2001**, *2*, 793–805. [CrossRef] [PubMed]

3. Ge, C.; Xiao, G.; Jiang, D.; Franceschi, R.T. Critical role of the extracellular signal-regulated kinase-MAPK pathway in osteoblast differentiation and skeletal development. *J. Cell Biol.* **2007**, *176*, 709–718. [CrossRef] [PubMed]

4. Xiao, G.; Jiang, D.; Gopalakrishnan, R.; Franceschi, R.T. Fibroblast growth factor 2 induction of the osteocalcin gene requires MAPK activity and phosphorylation of the osteoblast transcription factor, Cbfa1/Runx2. *J. Biol. Chem.* **2002**, *277*, 36181–36187. [CrossRef] [PubMed]

5. Rozario, T.; Desimone, D.W. The Extracellular Matrix In Development and Morphogenesis: A Dynamic View. *Dev. Biol.* **2011**, *341*, 126–140. [CrossRef] [PubMed]

6. Nakasaki, M.; Hwang, Y.; Xie, Y.; Kataria, S.; Gund, R.; Hajam, E.Y.; Samuel, R.; George, R.; Danda, D.; Paul, M.J.; et al. The matrix protein Fibulin-5 is at the interface of tissue stiffness and inflammation in fibrosis. *Nat. Commun.* **2015**, *6*, 8574. [CrossRef] [PubMed]

7. Aung, A.; Seo, Y.N.; Lu, S.; Wang, Y.; Jamora, C.; del Álamo, J.C. 3D Traction Stresses Activate Protease-Dependent Invasion of Cancer. *Biophys. J.* **2014**, *107*, 2528–2537. [CrossRef] [PubMed]

8. Varghese, S.; Elisseeff, J.H. Hydrogels for musculoskeletal tissue engineering. *Adv. Polym. Sci.* **2006**, *203*, 95–144.

9. Hersel, U.; Dahmen, C.; Kessler, H. RGD modified polymers: Biomaterials for stimulated cell adhesion and beyond. *Biomaterials* **2003**, *24*, 4385–4415. [CrossRef]

10. Musah, S.; Morin, S.; Wrighton, P.; Zwick, D.B.; Jin, S.; Kiessling, L.L. Glycosaminoglycan-binding hydrogels enable mechanical control of human pluripotent stem cell self-renewal. *ACS Nano* **2012**, *6*, 10168–10177. [CrossRef] [PubMed]

11. Elbert, D.L.; Hubbell, J.A. Surface Treatments of Polymers for Biocompatibility. *Annu. Rev. Mater. Sci.* **1996**, *26*, 365–394. [CrossRef]

12. Salinas, C.N.; Anseth, K.S. Mixed Mode Thiol-Acrylate Photopolymerizations for the Synthesis of PEG-Peptide Hydrogels. *Society* **2008**, *41*, 6019–6026. [CrossRef]

13. Hern, D.L.; Hubbell, J.A. Incorporation of adhesion peptides into nonadhesive hydrogels useful for tissue resurfacing. *J. Biomed. Mater. Res.* **1998**, *39*, 266–276. [CrossRef]

14. Dixon, J.E.; Shah, D.A.; Rogers, C.; Hall, S.; Weston, N.; Parmenter, C.D.J.; McNally, D.; Denning, C.; Shakesheff, K.M. Combined hydrogels that switch human pluripotent stem cells from self-renewal to differentiation. *Proc. Natl. Acad. Sci. USA* **2014**, *111*, 5580–5585. [CrossRef] [PubMed]

15. Rowley, J.A.; Madlambayan, G.; Mooney, D.J. Alginate hydrogels as synthetic extracellular matrix materials. *Biomaterials* **1999**, *20*, 45–53. [CrossRef]

16. Mann, B.K.; Gobin, A.S.; Tsai, A.T.; Schmedlen, R.H.; West, J.L. Smooth muscle cell growth in photopolymerized hydrogels with cell adhesive and proteolytically degradable domains: Synthetic ECM analogs for tissue engineering. *Biomaterials* **2001**, *22*, 3045–3051. [CrossRef]

17. Belair, D.G.; Le, N.N.; Murphy, W.L. Design of growth factor sequestering biomaterials. *Chem. Commun.* **2014**, *50*, 15651–15668. [CrossRef] [PubMed]

18. Battig, M.R.; Huang, Y.; Chen, N.; Wang, Y. Aptamer-functionalized superporous hydrogels for sequestration and release of growth factors regulated via molecular recognition. *Biomaterials* **2014**, *35*, 8040–8048. [CrossRef] [PubMed]

19. Jha, A.K.; Tharp, K.M.; Ye, J.; Santiago-Ortiz, J.L.; Jackson, W.M.; Stahl, A.; Schaffer, D.V.; Yeghiazarians, Y.; Healy, K.E. Enhanced survival and engraftment of transplanted stem cells using growth factor sequestering hydrogels. *Biomaterials* **2015**, *47*, 1–12. [CrossRef] [PubMed]

20. Cai, S.; Liu, Y.; Shu, X.Z.; Prestwich, G.D. Injectable glycosaminoglycan hydrogels for controlled release of human basic fibroblast growth factor. *Biomaterials* **2005**, *26*, 6054–6067. [CrossRef] [PubMed]

21. Chang, C.W.; Hwang, Y.; Brafman, D.; Hagan, T.; Dhung, C.; Varghese, S. Engineering cell-material interfaces for long-term expansion of human pluripotent stem cells. *Biomaterials* **2013**, *34*, 912–921. [CrossRef] [PubMed]

22. Sangaj, N.; Kyriakakis, P.; Yang, D.; Chang, C.-W.; Arya, G.; Varghese, S. Heparin mimicking polymer promotes myogenic differentiation of muscle progenitor cells. *Biomacromolecules* **2010**, *11*, 3294–3300. [CrossRef] [PubMed]

23. Benoit, D.S.W.; Durney, A.R.; Anseth, K.S. The effect of heparin-functionalized PEG hydrogels on three-dimensional human mesenchymal stem cell osteogenic differentiation. *Biomaterials* **2007**, *28*, 66–77. [CrossRef] [PubMed]

24. Lee, J.S.; Suarez-Gonzalez, D.; Murphy, W.L. Mineral coatings for temporally controlled delivery of multiple proteins. *Adv. Mater.* **2011**, *23*, 4279–4284. [CrossRef] [PubMed]

25. Steinmetz, N.J.; Bryant, S.J. Chondroitin sulfate and dynamic loading alter chondrogenesis of human mscs in peg hydrogels. *Biotechnol. Bioeng.* **2012**, *109*, 2671–2682. [CrossRef] [PubMed]

26. Ramaswamy, S.; Wang, D.-A.; Fishbein, K.W.; Elisseeff, J.H.; Spencer, R.G. An analysis of the integration between articular cartilage and nondegradable hydrogel using magnetic resonance imaging. *J. Biomed. Mater. Res. B Appl. Biomater.* **2006**, *77*, 144–148. [CrossRef] [PubMed]

27. Varghese, S.; Hwang, N.S.; Canver, A.C.; Theprungsirikul, P.; Lin, D.W.; Eliseeff, J. Chondroitin sulfate based niches for chondrogenic differentiation of mesenchymal stem cells. *Matrix Biol.* **2008**, *27*, 12–21. [CrossRef] [PubMed]

28. Chung, C.; Burdick, J.A. Influence of 3D Hyaluronic Acid Microenvironments on Mesenchymal Stem Cell Chondrogenesis. *Tissue Eng.* **2009**, *15*, 243–254. [CrossRef] [PubMed]

29. Culty, M.; Nguyen, H.A.; Underhill, C.B. The hyaluronan receptor (CD44) participates in the uptake and degradation of hyaluronan. *J. Cell Biol.* **1992**, *116*, 1055–1062. [CrossRef] [PubMed]

30. Burdick, J.A.; Prestwich, G.D. Hyaluronic acid hydrogels for biomedical applications. *Adv. Mater.* **2011**. [CrossRef] [PubMed]

31. Hlady, V.; Buijs, J. Protein adsorption on solid surfaces. *Curr. Opin. Biotechnol.* **1996**, *7*, 72–77. [CrossRef]

32. Ostuni, E.; Grzybowski, B.A.; Mrksich, M.; Roberts, C.S.; Whitesides, G.M. Adsorption of proteins to hydrophobic sites on mixed self-assembled monolayers. *Langmuir* **2003**, *19*, 1861–1872. [CrossRef]

33. Ayala, R.; Zhang, C.; Yang, D.; Hwang, Y.; Aung, A.; Shroff, S.S.; Arce, F.T.; Lal, R.; Arya, G.; Varghese, S. Engineering the cell-material interface for controlling stem cell adhesion, migration, and differentiation. *Biomaterials* **2011**, *32*, 3700–3711. [CrossRef] [PubMed]

34. Phillips, J.E.; Petrie, T.A.; Creighton, F.P.; García, A.J. Human mesenchymal stem cell differentiation on self-assembled monolayers presenting different surface chemistries. *Acta Biomater.* **2010**, *6*, 12–20. [CrossRef] [PubMed]

35. Benoit, D.S.W.; Schwartz, M.P.; Durney, A.R.; Anseth, K.S. Small functional groups for controlled differentiation of hydrogel-encapsulated human mesenchymal stem cells. *Nat. Mater.* **2008**, *7*, 816–823. [CrossRef] [PubMed]

36. Song, J.; Saiz, E.; Bertozzi, C.R. A New Approach to Mineralization of Biocompatible Hydrogel Scaffolds: An Efficient Process toward 3-Dimensional Bonelike Composites. *J. Am. Chem. Soc.* **2003**, *125*, 1236–1243. [CrossRef] [PubMed]

37. Suarez-Gonzalez, D.; Barnhart, K.; Migneco, F.; Flanagan, C.; Hollister, S.J.; Murphy, W.L. Controllable mineral coatings on PCL scaffolds as carriers for growth factor release. *Biomaterials* **2012**, *33*, 713–721. [CrossRef] [PubMed]

38. Phadke, A.; Zhang, C.; Hwang, Y.; Vecchio, K.; Vaghese, S. Templated mineralization of synthetic hydrogels for bone-like composite materials: Role of matrix hydrophobicity. *Biomacromolecules* **2010**, *11*, 2060–2068. [CrossRef] [PubMed]

39. Phadke, A.; Shih, Y.R.V.; Varghese, S. Mineralized Synthetic Matrices as an Instructive Microenvironment for Osteogenic Differentiation of Human Mesenchymal Stem Cells. *Macromol. Biosci.* **2012**, *12*, 1022–1032. [CrossRef] [PubMed]

40. Kang, H.; Shih, Y.-R.V.; Hwang, Y.; Wen, C.; Rao, V.; Seo, T.; Varghese, S. Mineralized gelatin methacrylate-based matrices induce osteogenic differentiation of human induced pluripotent stem cells. *Acta Biomater.* **2014**, *10*, 4961–4970. [CrossRef] [PubMed]

41. Kang, H.; Wen, C.; Hwang, Y.; Shih, Y.-R.V.; Kar, M.; Seo, S.W.; Varghese, S. Biomineralized matrix-assisted osteogenic differentiation of human embryonic stem cells. *J. Mater. Chem. B Mater. Biol. Med.* **2014**, *2*, 5676–5688. [CrossRef] [PubMed]

42. Shih, Y.-R.V.; Hwang, Y.; Phadke, A.; Kang, H.; Hwang, N.S.; Caro, E.J.; Nguyen, S.; Siu, M.; Theodorakis, E.A.; Gianneschi, N.C.; et al. Calcium phosphate-bearing matrices induce osteogenic differentiation of stem cells through adenosine signaling. *Proc. Natl. Acad. Sci. USA* **2014**, *111*, 990–995. [CrossRef] [PubMed]

43. Kang, H.; Shih, Y.-R.V.; Varghese, S. Biomineralized Matrices Dominate Soluble Cues To Direct Osteogenic Differentiation of Human Mesenchymal Stem Cells through Adenosine Signaling. *Biomacromolecules* **2015**, *16*, 1050–1061. [CrossRef] [PubMed]

44. Rao, V.; Shih, Y.-R.V.; Kang, H.; Kabra, H.; Varghese, S. Adenosine Signaling Mediates Osteogenic Differentiation of Human Embryonic Stem Cells on Mineralized Matrices. *Front. Bioeng. Biotechnol.* **2015**, *3*, 1–10. [CrossRef] [PubMed]
45. Valamehr, B.; Jonas, S.J.; Polleux, J.; Qiao, R.; Guo, S.; Gschweng, E.H.; Stiles, B.; Kam, K.; Luo, T.-J.M.; Witte, O.N.; et al. Hydrophobic surfaces for enhanced differentiation of embryonic stem cell-derived embryoid bodies results. *Proc. Natl. Acad. Sci. USA* **2008**, *105*, 14459–14464. [CrossRef] [PubMed]
46. Mei, Y.; Saha, K.; Bogatyrev, S.R.; Yang, J.; Hook, A.L.; Kalcioglu, Z.I.; Cho, S.-W.; Mitalipova, M.; Pyzocha, N.; Rojas, F.; et al. Combinatorial development of biomaterials for clonal growth of human pluripotent stem cells. *Nat. Mater.* **2010**, *9*, 768–778. [CrossRef] [PubMed]
47. De Silva, R.M.P.; Mano, J.F.; Reis, R.L. Smart thermoresponsive coatings and surfaces for tissue engineering: Switching cell-material boundaries. *Trends Biotechnol.* **2007**, *25*, 577–583. [CrossRef] [PubMed]
48. Yang, J.; Yamato, M.; Kohno, C.; Nishimoto, A.; Sekine, H.; Fukai, F.; Okano, T. Cell sheet engineering: Recreating tissues without biodegradable scaffolds. *Biomaterials* **2005**, *26*, 6415–6422. [CrossRef] [PubMed]
49. Varghese, S.; Lele, A.K.; Mashelkar, R.A. Designing new thermoreversible gels by molecular tailoring of hydrophilic-hydrophobic interactions. *J. Chem. Phys.* **2000**, *112*, 3063–3070. [CrossRef]
50. Shimizu, T.; Yamato, M.; Kikuchi, A.; Okano, T. Cell sheet engineering for myocardial tissue reconstruction. *Biomaterials* **2003**, *24*, 2309–2316. [CrossRef]
51. Zhang, S.; Burda, J.E.; Anderson, M.A.; Zhao, Z.; Ao, Y.; Cheng, Y.; Sun, Y.; Deming, T.J.; Sofroniew, M.V. Thermoresponsive Copolypeptide Hydrogel Vehicles for Central Nervous System Cell Delivery. *ACS Biomater. Sci. Eng.* **2015**, *1*, 705–717. [CrossRef]
52. Lei, Y.; Schaffer, D.V. A fully defined and scalable 3D culture system for human pluripotent stem cell expansion and differentiation. *Proc. Natl. Acad. Sci. USA* **2013**, *110*, E5039–E5048. [CrossRef] [PubMed]
53. Lim, H.L.; Chuang, J.C.; Tran, T.; Aung, A.; Arya, G.; Varghese, S. Dynamic Electromechanical Hydrogel Matrices for Stem Cell Culture. *Adv. Funct. Mater.* **2011**, *21*, 387–393. [CrossRef] [PubMed]
54. Chen, C.S.; Mrksich, M.; Huang, S.; Whitesides, G.M.; Ingber, D.E. Micropatterned surfaces for control of cell shape, position, and function. *Biotechnol. Prog.* **1998**, *14*, 356–363. [CrossRef] [PubMed]
55. Guvendiren, M.; Burdick, J.A. The control of stem cell morphology and differentiation by hydrogel surface wrinkles. *Biomaterials* **2010**, *31*, 6511–6518. [CrossRef] [PubMed]
56. Théry, M. Micropatterning as a tool to decipher cell morphogenesis and functions. *J. Cell Sci.* **2010**, *123*, 4201–4213. [CrossRef] [PubMed]
57. Revzin, A.; Russell, R.J.; Yadavalli, V.K.; Koh, W.-G.; Deister, C.; Hile, D.D.; Mellott, M.B.; Pishko, M.V. Fabrication of poly(ethylene glycol) hydrogel microstructures using photolithography. *Langmuir* **2001**, *17*, 5440–5447. [CrossRef] [PubMed]
58. Miller, J.S.; Stevens, K.R.; Yang, M.T.; Baker, B.M.; Nguyen, D.-H.T.; Cohen, D.M.; Toro, E.; Chen, A.A.; Galie, P.A.; Yu, X.; et al. Rapid casting of patterned vascular networks for perfusable engineered three-dimensional tissues. *Nat. Mater.* **2012**, *11*, 768–774. [CrossRef] [PubMed]
59. Cosson, S.; Lutolf, M.P. Hydrogel microfluidics for the patterning of pluripotent stem cells. *Sci. Rep.* **2014**, *4*, 4462. [CrossRef] [PubMed]
60. Martin, B.D.; Brandow, S.L.; Dressick, W.J.; Schull, T.L. Fabrication and application of hydrogel stampers for physisorptive microcontact printing. *Langmuir* **2000**, *16*, 9944–9946. [CrossRef]
61. Alom Ruiz, S.; Chen, C.S. Microcontact printing: A tool to pattern. *Soft Matter* **2007**, *3*, 168. [CrossRef]
62. Yeh, J.; Ling, Y.; Karp, J.M.; Gantz, J.; Chandawarkar, A.; Eng, G.; Blumling, J.; Langer, R.; Khademhosseini, A. Micromolding of shape-controlled, harvestable cell-laden hydrogels. *Biomaterials* **2006**, *27*, 5391–5398. [CrossRef] [PubMed]
63. Ji, Y.; Ghosh, K.; Shu, X.Z.; Li, B.; Sokolov, J.C.; Prestwich, G.D.; Clark, R.A.F.; Rafailovich, M.H. Electrospun three-dimensional hyaluronic acid nanofibrous scaffolds. *Biomaterials* **2006**, *27*, 3782–3792. [CrossRef] [PubMed]
64. Kim, J.; Yoon, J.; Hayward, R.C. Dynamic display of biomolecular patterns through an elastic creasing instability of stimuli-responsive hydrogels. *Nat. Mater.* **2010**, *9*, 159–164. [CrossRef] [PubMed]
65. McBeath, R.; Pirone, D.M.; Nelson, C.M.; Bhadriraju, K.; Chen, C.S. Cell shape, cytoskeletal tension, and RhoA regulate stem cell lineage commitment. *Dev. Cell* **2004**, *6*, 483–495. [CrossRef]
66. Kilian, K.A.; Bugarija, B.; Lahn, B.T.; Mrksich, M. Geometric cues for directing the differentiation of mesenchymal stem cells. *Proc. Natl. Acad. Sci. USA* **2010**, *107*, 4872–4877. [CrossRef] [PubMed]

67. Lee, J.; Abdeen, A.A.; Zhang, D.; Kilian, K.A. Directing stem cell fate on hydrogel substrates by controlling cell geometry, matrix mechanics and adhesion ligand composition. *Biomaterials* **2013**, *34*, 8140–8148. [CrossRef] [PubMed]

68. Lee, J.; Abdeen, A.A.; Huang, T.H.; Kilian, K.A. Controlling cell geometry on substrates of variable stiffness can tune the degree of osteogenesis in human mesenchymal stem cells. *J. Mech. Behav. Biomed. Mater.* **2014**, *38*, 209–218. [CrossRef] [PubMed]

69. Choi, C.H.; Hagvall, S.H.; Wu, B.M.; Dunn, J.C.Y.; Beygui, R.E.; C.J. Kim, C.J. Cell interaction with three-dimensional sharp-tip nanotopography. *Biomaterials* **2007**, *28*, 1672–1679. [CrossRef] [PubMed]

70. Dalton, B.A.; Walboomers, X.F.; Dziegielewski, M.; Evans, M.D.; Taylor, S.; Jansen, J.A.; Steele, J.G. Modulation of epithelial tissue and cell migration by microgrooves. *J. Biomed. Mater. Res.* **2001**, *56*, 195–207. [CrossRef]

71. Foley, J.D.; Grunwald, E.W.; Nealey, P.F.; Murphy, C.J. Cooperative modulation of neuritogenesis by PC12 cells by topography and nerve growth factor. *Biomaterials* **2005**, *26*, 3639–3644. [CrossRef] [PubMed]

72. Yim, E.K.F.; Pang, S.W.; Leong, K.W. Synthetic nanostructures inducing differentiation of human mesenchymal stem cells into neuronal lineage. *Exp. Cell Res.* **2007**, *313*, 1820–1829. [CrossRef] [PubMed]

73. McMurray, R.J.; Gadegaard, N.; Tsimbouri, P.M.; Burgess, K.V.; McNamara, L.E.; Tare, R.; Murawski, K.; Kingham, E.; Oreffo, R.O.C.; Dalby, M.J. Nanoscale surfaces for the long-term maintenance of mesenchymal stem cell phenotype and multipotency. *Nat. Mater.* **2011**. [CrossRef] [PubMed]

74. Dalby, M.J.; Gadegaard, N.; Tare, R.; Andar, A.; Riehle, M.O.; Herzyk, P.; Wilkinson, C.D.W.; Oreffo, R.O.C. The control of human mesenchymal cell differentiation using nanoscale symmetry and disorder. 2007. [CrossRef]

75. Lin, S.; Sangaj, N.; Razafiarison, T.; Zhang, C.; Varghese, S. Influence of physical properties of biomaterials on cellular behavior. *Pharm. Res.* **2011**, *28*, 1422–1430. [CrossRef] [PubMed]

76. Zhang, C.; Aung, A.; Liao, L.; Varghese, S. A novel single precursor-based biodegradable hydrogel with enhanced mechanical properties. *Soft Matter* **2009**, *5*, 3831. [CrossRef]

77. Chang, C.-W.; van Spreeuwel, A.; Zhang, C.; Varghese, S. PEG/clay nanocomposite hydrogel: A mechanically robust tissue engineering scaffold. *Soft Matter* **2010**, *6*, 5157. [CrossRef]

78. Haraguchi, K.; Takehisa, T. Nanocomposite hydrogels: A unique organic-inorganic network structure with extraordinary mechanical, optical, and swelling/De-swelling properties. *Adv. Mater.* **2002**, *14*, 1120–1124. [CrossRef]

79. Gong, J.P.; Katsuyama, Y.; Kurokawa, T.; Osada, Y. Double-network hydrogels with extremely high mechanical strength. *Adv. Mater.* **2003**, *15*, 1155–1158. [CrossRef]

80. Sun, J.-Y.; Zhao, X.; Illeperuma, W.R.K.; Chaudhuri, O.; Oh, K.H.; Mooney, D.J.; Vlassak, J.J.; Suo, Z. Highly stretchable and tough hydrogels. *Nature* **2012**, *489*, 133–136. [CrossRef] [PubMed]

81. Wozniak, M.A.; Chen, C.S. Mechanotransduction in development: A growing role for contractility. *Nat. Rev. Mol. Cell Biol.* **2009**, *10*, 34–43. [CrossRef] [PubMed]

82. Lo, C.M.; Wang, H.B.; Dembo, M.; Wang, Y.L. Cell movement is guided by the rigidity of the substrate. *Biophys. J.* **2000**, *79*, 144–152. [CrossRef]

83. Pelham, R.J.; Wang, Y.-L. Cell locomotion and focal adhesions are regulated by substrate flexibility. *Proc. Natl. Acad. Sci. USA* **1997**, *94*, 13661–13665. [CrossRef] [PubMed]

84. Banerjee, A.; Arha, M.; Choudhary, S.; Ashton, R.S.; Bhatia, S.R.; Schaffer, D.V.; Kane, R.S. The influence of hydrogel modulus on the proliferation and differentiation of encapsulated neural stem cells. *Biomaterials* **2009**. [CrossRef] [PubMed]

85. Engler, A.J.; Sen, S.; Sweeney, H.L.; Discher, D.E. Matrix Elasticity Directs Stem Cell Lineage Specification. *Cell* **2006**, *126*, 677–689. [CrossRef] [PubMed]

86. Ehrbar, M.; Sala, A.; Lienemann, P. Elucidating the Role of Matrix Stiffness in 3D Cell Migration and Remodeling. *Biophys. J.* **2011**, *100*, 284–293. [CrossRef] [PubMed]

87. McDaniel, D.P.; Shaw, G.A.; Elliott, J.T.; Bhadriraju, K.; Meuse, C.; Chung, K.-H.; Plant, A.L. The Stiffness of Collagen Fibrils Influences Vascular Smooth Muscle Cell Phenotype. *Biophys. J.* **2007**, *92*, 1759–1769. [CrossRef] [PubMed]

88. Hansen, L.K.; Wilhelm, J.; Fassett, J.T. Regulation of Hepatocyte Cell Cycle Progression and Differentiation by Type I Collagen Structure. *Curr. Top. Dev. Biol.* **2005**, *72*, 205–236.

89. Huebsch, N.; Arany, P.R.; Mao, A.S.; Shvartsman, D.; Ali, O.A.; Bencherif, S.A.; Rivera-Feliciano, J.; Mooney, D.J. Harnessing traction-mediated manipulation of the cell/matrix interface to control stem-cell fate. *Nat. Mater.* **2010**, *9*, 518–526. [CrossRef] [PubMed]

90. Khetan, S.; Guvendiren, M.; Legant, W.R.; Cohen, D.M.; Chen, C.S.; Burdick, J.A. Degradation-mediated cellular traction directs stem cell fate in covalently crosslinked three-dimensional hydrogels. *Nat. Mater.* **2013**, *12*, 458–465. [CrossRef] [PubMed]

91. Chaudhuri, O.; Gu, L.; Darnell, M.; Klumpers, D.; Bencherif, S.A.; Weaver, J.C.; Huebsch, N.; Mooney, D.J. Substrate stress relaxation regulates cell spreading. *Nat. Commun.* **2015**, *6*, 6364. [CrossRef] [PubMed]

92. Chaudhuri, O.; Gu, L.; Klumpers, D.; Darnell, M.; Bencherif, S.A.; Weaver, J.C.; Huebsch, N.; Lee, H.-P.; Lippens, E.; Duda, G.N.; et al. Hydrogels with tunable stress relaxation regulate stem cell fate and activity. *Nat. Mater.* **2015**. [CrossRef] [PubMed]

93. Lian, I.; Kim, J.; Okazawa, H.; Zhao, J.; Zhao, B.; Yu, J.; Chinnaiyan, A.; Israel, M.A.; Goldstein, L.S.B.; Abujarour, R.; et al. The role of YAP transcription coactivator in regulating stem cell self-renewal and differentiation. *Genes Dev.* **2010**, *24*, 1106–1118. [CrossRef] [PubMed]

94. Dupont, S.; Morsut, L.; Aragona, M.; Enzo, E.; Giulitti, S.; Cordenonsi, M.; Zanconato, F.; Le Digabel, J.; Forcato, M.; Bicciato, S.; et al. Role of YAP/TAZ in mechanotransduction. *Nature* **2011**, *474*, 179–183. [CrossRef] [PubMed]

95. Stowers, R.S.; Allen, S.C.; Suggs, L.J. Dynamic phototuning of 3D hydrogel stiffness. *Proc. Natl. Acad. Sci. USA* **2015**, *112*, 1953–1958. [CrossRef] [PubMed]

96. Mabry, K.M.; Lawrence, R.L.; Anseth, K.S. Dynamic stiffening of poly(ethylene glycol)-based hydrogels to direct valvular interstitial cell phenotype in a three-dimensional environment. *Biomaterials* **2015**, *49*, 47–56. [CrossRef] [PubMed]

97. Kloxin, A.M.; Kasko, A.A.; Salinas, C.N.; Anseth, K.S. Photodegradable Hydrogels for Dynamic Tuning of Physical and Chemical Properties. *Science* **2013**, *59*, 1–6. [CrossRef] [PubMed]

98. Guvendiren, M.; Burdick, J.A. Stiffening hydrogels to probe short- and long-term cellular responses to dynamic mechanics. *Nat. Commun.* **2012**, *3*, 792. [CrossRef] [PubMed]

99. Badiger, M.V.; McNeill, M.E.; Graham, N.B. Porogens in the preparation of microporous hydrogels based on poly(ethylene oxides). *Biomaterials* **1993**, *14*, 1059–1063.

100. Chiu, Y.-C.; Larson, J.C.; Isom, A.; Brey, E.M. Generation of porous poly(ethylene glycol) hydrogels by salt leaching. *Tissue Eng. C Methods* **2010**, *16*, 905–912. [CrossRef] [PubMed]

101. Hwang, Y.; Zhang, C.; Varghese, S. Poly(ethylene glycol) cryogels as potential cell scaffolds: Effect of polymerization conditions on cryogel microstructure and properties. *J. Mater. Chem.* **2010**, *20*, 345. [CrossRef]

102. Pham, Q.P.; Sharma, U.; Mikos, A.G. Electrospinning of polymeric nanofibers for tissue engineering applications: A review. *Tissue Eng.* **2006**, *12*, 1197–1211. [CrossRef] [PubMed]

103. Matthews, J.A.; Wnek, G.E.; Simpson, D.G.; Bowlin, G.L. Electrospinning of Collagen Nanofibers. *Biomacromolecules* **2002**, *3*, 232–238. [CrossRef] [PubMed]

104. Li, L.; Hsieh, Y.-L. Ultra-fine polyelectrolyte hydrogel fibres from poly(acrylic acid)/poly(vinyl alcohol). *Nanotechnology* **2005**, *16*, 2852–2860. [CrossRef]

105. Loh, Q.L.; Choong, C. Three-dimensional scaffolds for tissue engineering applications: Role of porosity and pore size. *Tissue Eng. B Rev.* **2013**, *19*, 485–502. [CrossRef] [PubMed]

106. Zeng, L.; Yao, Y.; Wang, D.A.; Chen, X. Effect of microcavitary alginate hydrogel with different pore sizes on chondrocyte culture for cartilage tissue engineering. *Mater. Sci. Eng. C* **2014**, *34*, 168–175. [CrossRef] [PubMed]

107. Oliviero, O.; Ventre, M.; Netti, P.A. Functional porous hydrogels to study angiogenesis under the effect of controlled release of vascular endothelial growth factor. *Acta Biomater.* **2012**, *8*, 3294–3301. [CrossRef] [PubMed]

108. Dziubla, T.D.; Lowman, A.M. Vascularization of PEG-grafted macroporous hydrogel sponges: A three-dimensional in vitro angiogenesis model using human microvascular endothelial cells. *J. Biomed. Mater. Res. A* **2004**, *68*, 603–614. [CrossRef] [PubMed]

109. Matsiko, A.; Gleeson, J.P.; O'Brien, F.J. Scaffold Mean Pore Size Influences Mesenchymal Stem Cell Chondrogenic Differentiation and Matrix Deposition. *Tissue Eng. A* **2015**, *21*, 486–497. [CrossRef] [PubMed]

110. Bakshi, A.; Fisher, O.; Dagci, T.; Himes, T.; Fischer, I.; Lownan, A. Mechanically engineered hydrogel scaffolds for axonal growth and angiogenesis after transplantation in spinal cord injury. *J. Neurosurg. Spine* **2004**, *1*, 322–329. [CrossRef] [PubMed]

111. Phadke, A.; Hwang, Y.; Kim, S.H.; Kim, S.H.; Yamaguchi, T.; Masuda, K.; Varghese, S. Effect of scaffold microarchitecture on osteogenic differentiation of human mesenchymal stem cells. *Eur. Cell Mater.* **2013**, *25*, 114–128. [PubMed]

112. Metters, A.T.; Anseth, K.S.; Bowman, C.N. Fundamental studies of a novel, biodegradable PEG-*b*-PLA hydrogel. *Polymer (Guildf.)* **2000**, *41*, 3993–4004. [CrossRef]

113. Alsberg, E.; Kong, H.J.; Hirano, Y.; Smith, M.K.; Albeiruti, A.; Mooney, D.J. Regulating bone formation via controlled scaffold degradation. *J. Dent. Res.* **2003**, *82*, 903–908. [CrossRef] [PubMed]

114. Temenoff, J.S.; Park, H.; Jabbari, E.; Conway, D.E.; Sheffield, T.L.; Ambrose, C.G.; Mikos, A.G. Thermally cross-linked oligo(poly(ethylene glycol) fumarate) hydrogels support osteogenic differentiation of encapsulated marrow stromal cells in vitro. *Biomacromolecules* **2004**, *5*, 5–10. [CrossRef] [PubMed]

115. Lutolf, M.P.; Lauer-Fields, J.L.; Schmoekel, H.G.; Metters, A.T.; Weber, F.E.; Fields, G.B.; Hubbell, J.A. Synthetic matrix metalloproteinase-sensitive hydrogels for the conduction of tissue regeneration: Engineering cell-invasion characteristics. *Proc. Natl. Acad. Sci. USA* **2003**, *100*, 5413–5418. [CrossRef] [PubMed]

116. Jo, Y.S.; Rizzi, S.C.; Ehrbar, M.; Weber, F.E.; Hubbell, J.A.; Lutolf, M.P. Biomimetic PEG hydrogels crosslinked with minimal plasmin-sensitive tri-amino acid peptides. *J. Biomed. Mater. Res. A* **2010**, *93*, 870–877. [CrossRef] [PubMed]

117. Aimetti, A.A.; Tibbitt, M.W.; Anseth, K.S. Human neutrophil elastase responsive delivery from poly (ethylene glycol) hydrogels. *Biomacromolecules* **2009**, *10*, 1484–1489. [CrossRef] [PubMed]

118. Kar, M.; Vernon Shih, Y.R.; Velez, D.O.; Cabrales, P.; Varghese, S. Poly(ethylene glycol) hydrogels with cell cleavable groups for autonomous cell delivery. *Biomaterials* **2016**, *77*, 186–197. [CrossRef] [PubMed]

*gels*

MDPI

*Review*

# Bioresponsive Hydrogels: Chemical Strategies and Perspectives in Tissue Engineering

**Antonella Sgambato, Laura Cipolla * and Laura Russo ***

Department of Biotechnology and Biosciences, University of Milano-Bicocca, 20126 Milano, Italy; antonella.sgambato@unimib.it
* Correspondence: laura.cipolla@unimib.it (L.C.); laura.russo@unimib.it (L.R.); Tel.: +39-264-483-460 (L.C.)

Academic Editor: Esmaiel Jabbari
Received: 4 August 2016; Accepted: 8 October 2016; Published: 14 October 2016

**Abstract:** Disease, trauma, and aging account for a significant number of clinical disorders. Regenerative medicine is emerging as a very promising therapeutic option. The design and development of new cell-customised biomaterials able to mimic extracellular matrix (ECM) functionalities represents one of the major strategies to control the cell fate and stimulate tissue regeneration. Recently, hydrogels have received a considerable interest for their use in the modulation and control of cell fate during the regeneration processes. Several synthetic bioresponsive hydrogels are being developed in order to facilitate cell-matrix and cell-cell interactions. In this review, new strategies and future perspectives of such synthetic cell microenvironments will be highlighted.

**Keywords:** hydrogels; tissue engineering; bioconjugation

## 1. Introduction

Bioresponsive hydrogels are dynamic systems that are capable of responding to or stimulating specific signals through the natural biological processes [1]. The possibility to tailor hydrogel composition, stiffness, and degradation rate makes this class of materials promising tools for tissue engineering applications. Hydrogels are cross-linked 3D networks containing hydrophilic polymer chains able to adsorb a significant amount of water [2] with a high versatility degree of chemical and physical properties. In addition, hydrogels in vivo administration can be performed by minimally invasive methods in order to avoid complex surgical intervention, even in irregular target sites of injured tissues [2]. Among these, bioresponsive hydrogels are considered "smart" biomaterials and are attracting great interest thanks to their controllable physical and biochemical properties when exposed to specific conditions in our body such as temperature, pH, or enzymes and receptors [3]. These general features may result in hydrogel responses (i.e., in term of swelling, degradation, mechanical deformation) [4] and/or in cells and tissue responses. This ability can be obtained and controlled by the spatial functionalization of hydrogel constituents with specific biological entities (biocues) in order to induce the desired stimuli [5]. Biological entities used with these aims can be native or synthetic biomacromolecules, such as enzymes, antibodies, nucleic acids [6], or small bioactive molecules such as carbohydrates [7] or peptides [8]. In this review, we will focus on a brief overview of the strategies employed to obtain bioresponsive hydrogel through functionalization with bioactive molecules.

## 2. Classification of the Hydrogels

On the basis of their composition, it is possible to distinguish natural, synthetic, and composite or hybrid hydrogels (Figure 1).

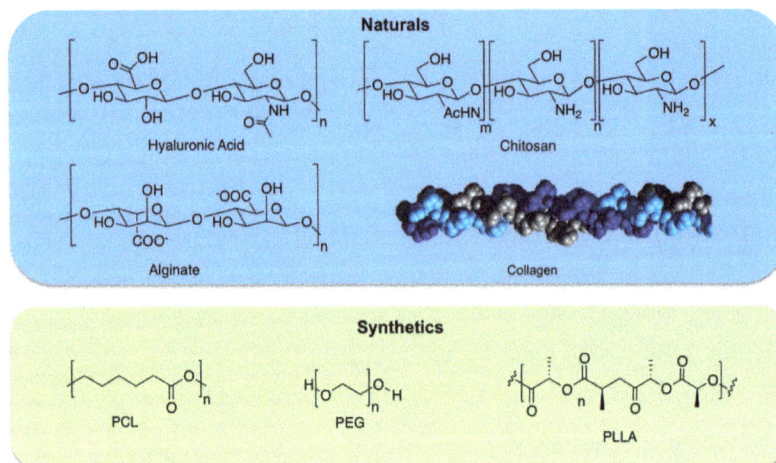

**Figure 1.** Chemical structure of some natural, and synthetic hydrogels.

*2.1. Natural Hydrogels*

Natural hydrogels are made up of natural biomacromolecules including proteins and polysaccharides. These biomacromolecules have different natural origins (animals, plants, or microorganisms). Several polysaccharide-based or protein-based hydrogels have been synthesized. A number of examples of polysaccharide-based hydrogels with tuneable properties are available for cartilage and bone tissue engineering applications. Most cited hydrogels for these applications have been prepared using hyaluronic acid [9], chitosan [10], and alginate [11] biopolymers. Hyaluronic acid (HA), alginate, and chitosan are established examples of polysaccharides used in tissue engineering applications. It is known that HA is a non-sulphated glycosaminoglycan (GAG) ubiquitous in mammalian tissues. It is a linear polysaccharide composed of a repeating disaccharide of (1–3) and (1–4)-linked β-D-glucuronic acid and N-acetyl-β-D-glucosamine units, overproduced during the wound healing process.

HA-based hydrogels have been produced by covalent cross-linking, for example, by hydrazide derivatives, by esterification, by carbodiimide chemistry, and by Huisgen-type cycloaddition (click chemistry) [12]. Additionally, HA has also been combined with synthetic polymers, proteins, and peptides in order to produce hybrid hydrogels [13]. Chitosan has been employed for different tissue regenerative approaches. It is a polysaccharide made up of (1–4)-linked D-glucosamine and N-acetyl-D-glucosamine units derived from chitin, a constituent of arthropod exoskeletons, but its structure is similar to naturally occurring GAGs. One of the main chitosan properties is its solubility in diluted acids by protonation of the amino groups so that it can be gelled, for example, by pH increase. Chitosan hydrogels were also obtained via UV-assisted photopolymerization, via cross-linking agents (i.e., genipin, glutaraldehyde, and squarate), or thermal variations [14].

Alginates are composed of guluronic acid and mannuronic acid. Their abundance, and low prices, allow a widespread use in the food industry as thickeners, emulsifiers, and in tissue engineering applications. Alginate was also used for various biomedical applications, such as drug delivery and cell encapsulation, because it is able to gel under mild conditions by the addition of divalent cations, it is biodegradable, and has low toxicity. Alginate-based hydrogels have also been obtained by chemical cross-linking by adipic hydrazide or PEG using carbodiimide chemistry [15].

Thanks to their structural role in nature, also protein-based hydrogels gained great interest for tissue engineering [16]. Collagen-based hydrogels have found application due to its ubiquitous presence in different tissues of the human body. Several cross-linking strategies were performed to control mechanical properties and 3D structures using for example 1,4-butanediol diglycidyl ether (BDDGE) or genipin cross-linking [17,18]. Another protein, gelatin (obtained from collagen hydrolysis) has captured increasing attention as it has relatively low antigenicity although maintaining the properties of biocompatibility and biodegradability; moreover, gelatin is significantly less expensive than collagen [19]. Several examples of photocrosslinked gelatin (i.e., gelatin methacrylamide (GelMA)) with tunable physico-chemical and biological properties were investigated for tissue engineering applications [20,21]. Other chemoselective cross-linking strategies involve thiol-ene photopolymerization between thiolated gelatin and pentenoyl gelatin [22] or thiolated gelatin and PEGdA [23]. These are just a few examples of natural biopolymers used for tissue engineering applications.

## 2.2. Synthetic Hydrogels

Even though hydrogels derived from natural biomacromolecules, such as proteins or polysaccharides, can actively support cell viability or cell differentiation, these materials have poor mechanical features and they are hard to process, and it is also difficult to maintain product consistency. To overcome these limitations, synthetic materials represent promising starting materials in scaffold design [24]. In fact, synthetic hydrogels can be designed using polymers with controlled molecular weight and biodegradable linkers. These features allow for the fine tuning of hydrogel composition, formation dynamics, mechanical properties, and degradation rates. Examples of synthetic materials are poly(ethylene glycol) (PEG), poly(lactide) (PLA), and poly($\varepsilon$-caprolactone) (PCL). PEG is a hydrophilic polymer, currently Food and Drug Administration (FDA) approved for several biomedical applications, and is one of the most frequently applied synthetic polymers for hydrogel preparation. Usually PEG acrylates or methacrylates can be photopolymerized, affording controlled hydrogel architectures [25]. Thermally reversible hydrogels have also been produced from block copolymers of the hydrophilic PEG and several polyesters as the hydrophobic block; PLA, poly(glycolic-*co*-lactic acid) (PLGA), PCL, and poly(L-lactic acid) (PLLA) are extensively used [26] because of their biocompatibility, biodegradability and facile synthesis by ring opening polymerization of lactide, glycolide, or $\varepsilon$-caprolactone monomers.

However, synthetic polymers lack cell adhesion sites and specific characteristics able to induce biological responses. In order to overcome these drawbacks, extensive conjugation chemistry has been applied to introduce bioactive molecules, such as the well-known RGD peptide (arginine-glycine-aspartic acid) to allow cell adhesion [27], or other molecules able to induce and guide specific biological phenomena (i.e., growth factors or carbohydrates able to guide cell differentiation) [28]. Different bioconjugation methods have been developed to introduce biological functions into synthetic materials [29].

## 3. From Tissue Complexity to Hydrogel Design: The Simplification Game

In the area of tissue engineering, hydrogels can be tuned in order to satisfy different design parameters for their integration in the damaged site, and, therefore, to operate in an appropriate manner and support new tissue formation. Hydrogel-based materials are able to supply a three-dimensional structure to adequately support cell proliferation, tissue integration, and differentiation. This tipology of 3D structure reproduces very closely the nature of tissues and takes into account the morphology and gene expression. In order to develop hydrogels for tissue engineering, different parameters must be considered, including mechanical and physico-chemical properties (i.e., biodegradation) and biological performance parameters (i.e., biocompatibility, cell adhesion, and proliferation). Moreover, it is important to consider the availability and commercial feasibility when producing hydrogels for tissue engineering applications. The native extracellular matrix (ECM) supplies a plethora of signals to neighbouring cells that modulate functional outputs, in combination with cell-cell and cell-ECM signalling [30]. Moreover, the extracellular matrix should be imagined as a dynamic milieu where the local environment can be reshaped through cell-mediated secretion and deposition of biomolecules or degraded through cell-secreted enzymes called matrix metalloproteinases (MMPs). In nature, the in vivo organization of tissues, with regard to their development and remodelling, is controlled by several factors that regulate and interplay at different levels in time and space. In tissue engineering and regenerative medicine studies, an adequate environment is necessary to supply the essential factors able to drive cell functions in vivo; in this way, it is possible to direct cells to differentiate in the right way [31]. Hydrogels made up of naturally derived components, for example extracellular matrix proteins (such as collagen) and polysaccharides like GAGs, have received particular attention in different applications in the field of regenerative medicine [5], since they supply physico-chemical and biochemical features that are similar to the native cellular milieu. Mimicking the natural tissue and microenvironment is not a simple objective; the simplification is possible thanks to the growth in the knowledge about signalling processes that regulate cell behaviour. To promote the understanding of the cell-niche interactions, hydrogels have been extensively employed as artificial cell niche given their tissue-like water content as well as tunable physico-chemical properties. However, only a few hydrogels produced to date allow the control of cell microenvironment properties such as biochemical signals and mechanical stiffness [32].

## 4. Bioactivation Strategies

Bio(macro)molecule conjugation (bioconjugation) is a useful approach that allows for the improvement of the hydrogels through signalling molecules, such as proteins, peptides [33,34], or even carbohydrate epitopes [35–37], in order to modulate and drive cellular fate. The conjugated biomolecules can be chosen in order to mediate different cellular events that are fundamental in regenerative medicine approaches, such as adhesion, migration, proliferation, and differentiation. Probably, the idea of bioresponsive hydrogel came from the pivotal study by Tirrell and co-workers [38] on self-assembly and gelation of a triblock artificial protein that was conjugated to a PEG moiety in order to control the gelling properties as a function of pH and temperature. Generally speaking, bioconjugated hydrogels may be classified into different groups depending on the bioactive molecule used for bioresponse activation: hydrogel-protein conjugates, hydrogel-peptide conjugates, hydrogel–glycan conjugates, and hydrogel-small molecule conjugates.

### 4.1. Hydrogel–Protein Conjugates

Proteins are fundamental and ubiquitous macromolecules playing key roles in the body mediating a plethora of cell processes and, for this reason, relevant for the design of hydrogel for tissue engineering approaches [39,40]. The incorporation of proteins and bioactive molecules in hydrogels allows for the control of the cellular microenvironment, and may be a valuable tool to support the regenerative process. However, their complex structure and, in some cases, the limited availability through biotechnology and molecular biology techniques may hamper the development of protein-conjugated hydrogels. Moreover, protein may have poor chemical stability and, together with their tendency to aggregate, their bioactivity may be limited. Despite these issues, various approaches have been developed to incorporate and deliver proteins into hydrogels. In general, different strategies can be used to include proteins into hydrogels, such as physical entrapment or covalent linkage to the hydrogel macromolecules. The bioconjugation through covalent bonds may offer some advantages in terms of stability against in vivo degradation forces and the maintenance of the therapeutic concentration in the hydrogel. A few examples of smart strategies for covalent protein bioconjugated hydrogels will be given in this section.

Anseth and co-workers fabricated PEG hydrogels via a thiol-acrylate photopolymerization reaction; in particular, PEG-diacrylate precursors were conjugated to thiolated EphA5-Fc receptor and ephrinA5-Fc ligand for enhancing pancreatic β-cell survival (Figure 2) [41]. EphA receptor and ephrinA ligand are cell surface-bound proteins involved in, among other things, insulin secretion from pancreatic β-cells promoting cell adhesion and motility/morphology changes through the integrin signalling pathway. These bioresponsive hydrogels were shown to provide crucial cell-cell communication cues for cell survival and proliferation.

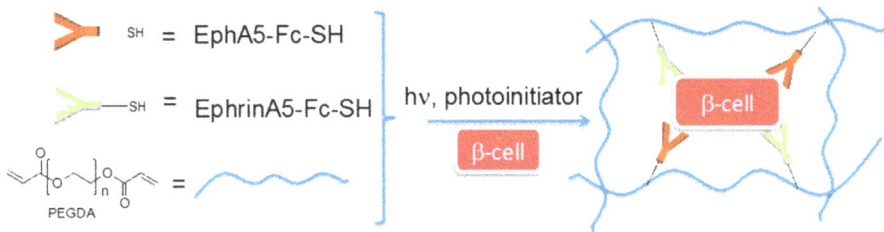

**Figure 2.** Schematic representation of poly(ethylene glycol) (PEG)-based hydrogels conjugated to EphA5-Fc and EphrinA5-Fc, tailored to sustain pancreatic β-cells survival.

In other examples, transforming growth factor-β1 (TGF-β1) was linked to photopolymerizable methacrylated chitosan (MeGC) hydrogels [42] and to acrylated HA hydrogels [43]. The first hydrogel was successfully used to induce chondrogenesis in human mesenchymal stem cells (hMSCs), while the second one showed enhanced cell survival and engraftment of encapsulated murine cardiac progenitors to the host tissue after transplantation, accompanied by vascularization.

Toward cardiac tissue regeneration, several hydrogels functionalised with different or multiple growth factors have been proposed. For example, stromal derived factor-1α (SDF-1α) conjugated to a PEG hydrogel [44] in vivo demonstrated a sustained colonization of progenitor cells in the ischemic tissue and promoted the angiogenetic process. Similarly, the recruitment of progenitor cells induced re-epithelization and revascularization with alginate-based hydrogels bioconjugated to SDF-1α [45]. Another hot issue in regenerative medicine is neural tissue regeneration [46,47]. In this respect, an increase of axon outgrowth of dopaminergic neurons, from rat embryos or differentiated from stem cells in culture, was obtained through an injectable PEG-silica composite hydrogel conjugated to semaphorin 3A. Semaphorins are a class of secreted and membrane proteins particularly relevant in neural system development; they are involved in axonal growth cone guidance and in the deflection of axons from inappropriate regions. However, the presence of non-degradable silica particles embedded in the hydrogel resulted in an increase of macrophages and glial cells in long term implantation studies.

In general, the chemical bioconjugation of complex signalling proteins is a hard task, since the activity and stability of proteins can be compromised by the conjugation reaction that might not be sufficiently mild and/or chemoselective. In order to address this issue, an extremely interesting approach was proposed by Lutolf and co-workers, who obtained the spatiotemporally controlled enzyme-mediated bioconjugation of vascular endothelial growth factor 121 (VEGF$_{121}$) and the recombinant fibronectin type III repeat 9-10 fragment (FN$_{9-10}$) [48]. Briefly, the researchers synthesised a PEG-based hydrogel functionalised with a "masked" enzyme substrate that could be rendered accessible to the related enzyme by a photocatalysed reaction (Figure 3). Once the substrate is accessible, the enzyme (that is FXIIIa, a key enzyme involved in the blood coagulation cascade) catalyses the reaction between the ε-amino group of lysine (K) and the carboxyamide side chain of glutamine (Q) to the corresponding ε-(γ-glutamyl)lysine isopeptide. Using recombinant VEGF$_{121}$ engineered with an exogenous peptide domain at the *N*-terminus containing the required glutamine to enable enzymatic cross-linking (Q-peptide in Figure 3), the enzymatic reaction affords the PEG hydrogel conjugated to the growth factor. The patterned PEG hydrogel is biocompatible for mesenchymal stem cells (MSC) cells.

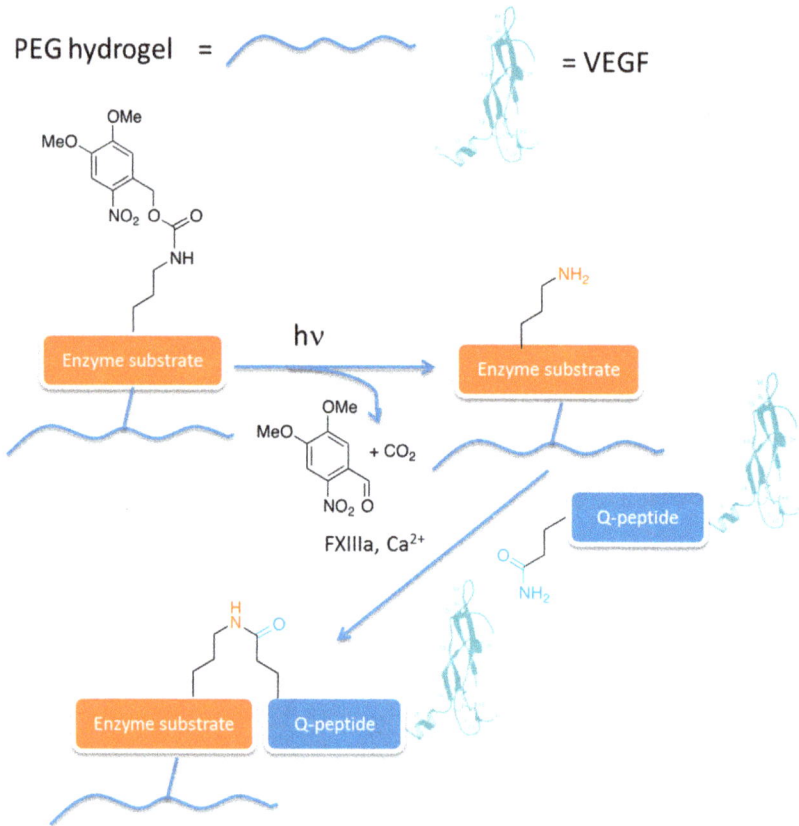

**Figure 3.** Strategy for the synthesis of PEG hydrogels conjugated to vascular endothelial growth factor (VEGF), through an enzyme-catalysed bioconjugation step, proposed to support primary human mesenchymal stem cells growth.

### 4.2. Hydrogel–Peptide Conjugates

The bioconjugation of peptides in place of full-length proteins greatly simplifies the chemistry and may improve the efficacy of the bioconjugation. Since many mammalian cells are anchorage-dependent, cell adhesive properties of the hydrogel can be achieved by the introduction of small adhesive peptidic sequences into the hydrogel matrix [49]; for example, they can be derived from laminin, such as RGD (Arg-Gly-Asp) [50], LGTIPG, YIGSR, IKVAV, LRE, PDGSR, IKLLI, LRGDN [51], and from type I collagen and fibronectin, in other words, DGEA [52], KQAGDV, REDV, and PHSRN. PEG-based hydrogels are useful for tissue engineering applications due to their favourable porosity, mechanical properties, and biocompatibility. However, due to their chemical nature, PEG does not possess cell attachment motifs. The conjugation of adhesive peptides such as the RGD sequence improves PEG hydrogel biofunctionality, rendering PEG-based hydrogels more suitable mimics of ECM. In this respect, Anseth and co-workers synthesised a dynamically controlled PEG-based hydrogel containing the adhesive sequence RGD through a double-click reaction (Figure 4) [53].

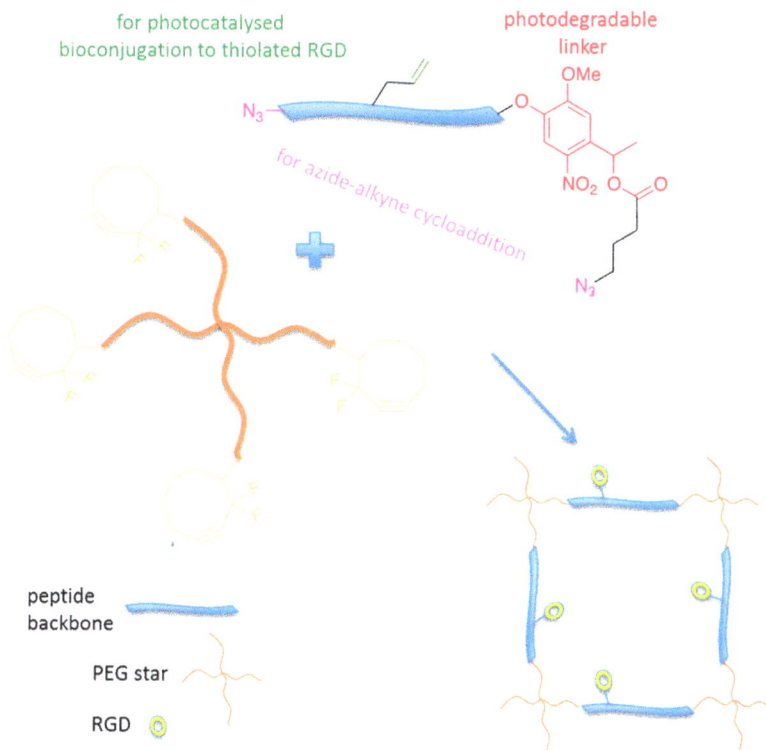

**Figure 4.** Strategy for the synthesis of photodegradable adhesive hydrogels for advanced 3D cell culture.

The spacer length between adhesive peptides and the material scaffold may be a critical parameter for regulating cell phenotype in tissue engineering. An interesting example of this issue was given with alginate hydrogels functionalised with RGD peptides linked through varying spacer arm lengths and assessed with primary human fibroblasts either on 2D or 3D scaffolds [54]. Alginate hydrogels were functionalised with G$_n$RGDSP moieties (n = number of glycine units) through EDC/sulfo-NHS coupling. Four glycine units in the spacer arm is essential for enhanced adhesion and growth of fibroblasts. An optimal spacer length was also needed for minimizing cellular stress, as determined by the expression of heat shock proteins and Bcl-2. Despite the extensive use of adhesive sequences, such as RGD and close peptides, other amino acid sequences have been conjugated to hydrogels in order to stimulate different cellular responses. For example, in order to control and reduce the oxidative stress experienced by cardiomiocytes (CMs) in myocardial infarction, the antioxidant tripeptide glutathione was bioconjugated to a chitosan-based hydrogel (Figure 5) [55]. The glutathione-conjugated chitosan is effective in vitro as an antioxidant, is biocompatible in the presence of cardiac myocytes, and is able to suppress the oxidative damage and apoptosis in CMs by removing the excessive intracellular reactive oxygen species (ROS) content.

Still in cardiac tissue engineering, the peptide sequence QHREDGS, derived from the fibrinogen-like domain of angiopoietin-1, was conjugated to a collagen-chitosan hydrogel. Peptide-modified hydrogels induce the tube-like structure formation in encapsulated endothelial cells [56].

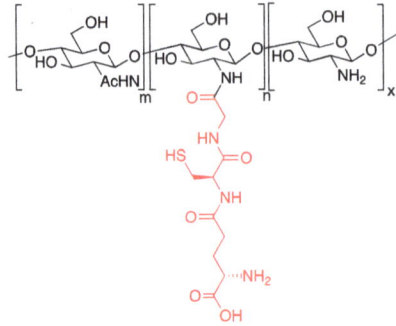

**Figure 5.** The chitosan-glutathione conjugated hydrogel able to suppress oxidative stress in cardiomiocytes.

In order to promote osteogenesis, alginate hydrogels bioconjugated to peptide mimics of BMP-2 were studied with encapsulated osteoblasts and mesenchymal stem cells. Two different peptide sequences derived from bone morphogenetic protein 2 (BMP-2) were incorporated into the alginate backbone by two different chemoselective strategies that could guarantee the spatial orientation of the peptides in their active form. The peptide DWIVA was covalently bound to alginate by carbodiimide chemistry through a four glycine spacer (Figure 6a) [57], while the so-called "knuckle epitope" of BMP-2 (KIPKASSVPTELSAISTLYL), was modified with a cysteine residue at the *N*-terminal end; the thiol group was conjugated via a Michael addition to the alginate modified with maleimide groups (Figure 6b).

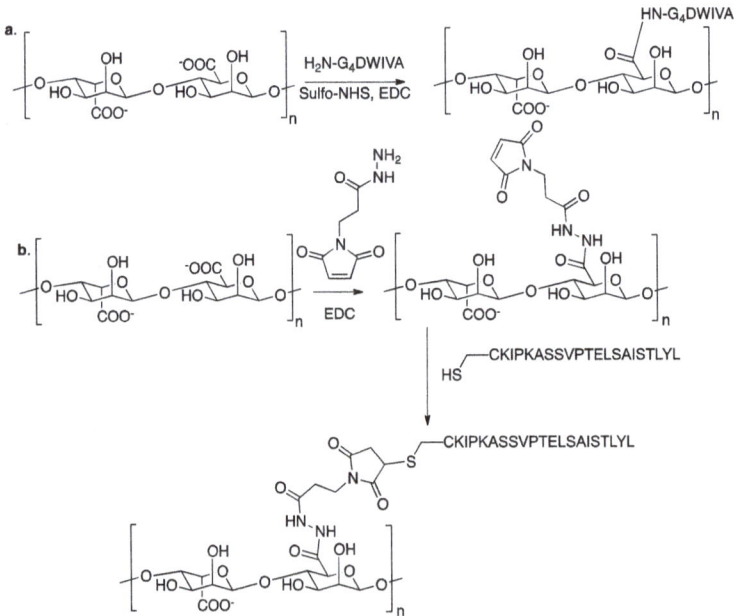

**Figure 6.** Alginate-based hydrogel designed to promote osteogenesis in murine mesenchymal stem cells. (**a**) Conjugation strategy for the DWIVA peptide; (**b**) Conjugation strategy for the "knuckle epitope".

Alginate functionalised with the knuckle epitope was shown to increase alkaline phosphatase activity in clonally derived murine osteoblasts, while with clonally derived murine mesenchymal stem cells it initiated Smad signalling, up-regulated osteopontin production, and increased mineral deposition.

Alginate-based hydrogels were also bioconjugated to the YIGSR peptide through amide bonds to the carboxylic acid groups of the alginate. The peptide-modified alginate hydrogels allowed adhesion of NB2a neuroblastoma cells and promoted neurite outgrowth [58]. The adhesion of NB2a neuroblastoma cells and neurite outgrowth was found to be a function of the peptide density.

*4.3. Hydrogel–Glycan Conjugates*

Thanks to their unique properties in signalling and cell development, glycans represent an interesting class of molecules to incorporate in hydrogels to make them bioresponsive. There are several examples of glycan epitopes used to bio-activate natural and synthetic materials [7]. In hydrogel systems glycan conjugates have been used in order to control both cell growth and vitality [59,60]. Heparin possesses binding domains to many growth factors, hence, when included in hydrogel design it may be a useful tool to allow for the retention and subsequent delivery of growth factors [59]. This approach was used to promote angiogenesis. In more detail, tyramine was first introduced into the gelatin backbone as the cross-linking points, then heparin was covalently linked to gelatin-tyramine. Vascular endothelial growth factor (VEGF) was then incorporated into the gelatin derivative by non-bonding interactions with heparin binding motifs and finally enzymatic reactions with hydrogen peroxide ($H_2O_2$) and horseradish peroxidase (HRP) allowed the formation of the gel by oxidative cross-linking [61]. In vivo implantation experiments showed deeper and denser cell infiltration and angiogenesis in the heparin-modified gelatin/VEGF gels than in the controls.

Long-term in vitro maintenance of primary hepatocytes is a burden for hepatic tissue engineering, since these cells lose their phenotype in standard culture conditions. Foster et al. studied the use of heparin-containing hydrogels as scaffolds for the culture and for the maintenance of functional primary hepatocytes [60]. Thiolated heparin was conjugated to diacrylated PEG via a thiol-ene photocatalysed reaction. Analysis of hepatic functionality of rat hepatocytes cultured on the hydrogel revealed that cells sustained albumin secretion for at least three weeks and increased cytochrome P450 activity. In addition, hepatocyte growth factor (HGF) was also entrapped into the gels thanks to heparin interactions; in the presence of HGF, higher amounts of albumin could be observed.

*4.4. Hydrogel–Small Molecule Conjugates*

Promising approaches in tissue regeneration rely on both suitable scaffold design and on the application of stem cells and the development of efficient approaches to control their fate. Recently, several small molecules have been identified as able to induce stem cell differentiation both in vitro and in vivo [62]. Within this frame, several hydrogels conjugated to small molecules, either to promote gelation or to impart selected biological stimuli, have been synthesised.

For example, tyramine [63,64] and dopamine [65] conjugated alginate have been prepared in order to induce gelation through enzymatic oxidation; dopamine was conjugated to alginate (Figure 7) and subsequently gelled by enzymatic cross-linking mediated by horseradish peroxidase (HRP) and $H_2O_2$.

The resulting hydrogels are cytocompatible when assayed with NIH 3T3 cells. In addition, compared with unfunctionalised alginate hydrogels, the dopamine-conjugated hydrogel showed higher cell adhesion and elasticity properties.

Catechol moieties were also grafted onto chitosan by reductive amination, and the resulting macromers were gelled through coordinative interaction with transition metal ions, such as iron, or in oxidative conditions by $NaIO_4$ [66].

Chitosan–catechol conjugates were also used as precursors for the synthesis of hybrid hydrogels in the presence of thiolated Pluronic F-127 triblock copolymer; the resulting hybrid material showed temperature-sensitive and adhesive properties to mucous layers and to soft tissues, together with good hemostatic properties useful for wound healing [67].

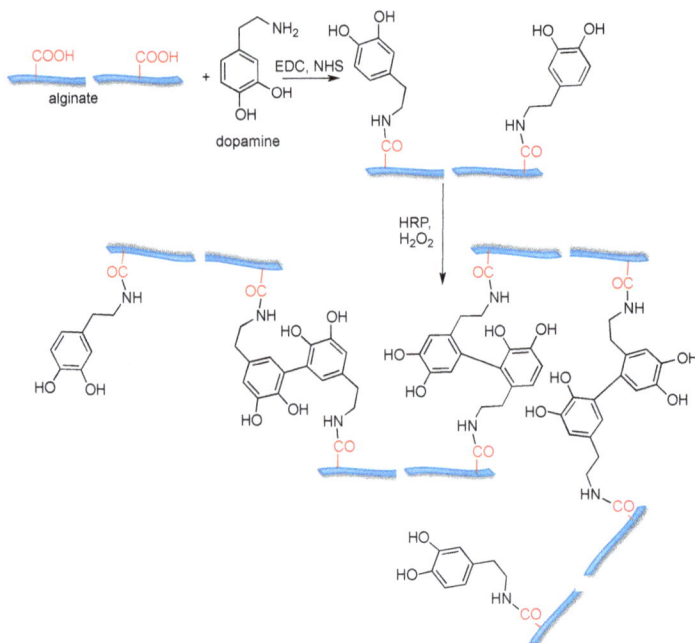

**Figure 7.** Alginate-based hydrogels cross-linked through dopamine oxidation by horseradish peroxidase (HRP); biocompatibility was assayed with NIH 3T3 cells.

Other small molecule motifs have been proposed for the formation of biofunctional supramolecular hydrogels, such as bisphosphonates or glucosamine, that showed wound-healing properties [68].

Small molecules can be grafted onto hydrogel macromers in order to induce a specific cellular response or to promote cell differentiation. Bisphosphonate moieties have been grafted onto hyaluronic acid hydrogels in order to promote BMP-2 sequestration and bone regeneration [69]. In more detail, hyaluronic acid precursors were covalently functionalised with bisphosphonate ligands; the BP moieties are efficient sequestering agents of BMP-2 that can be protected from degradation and released in the site of administration by enzymatic hydrolysis by hyaluronidases.

Biological studies showed that BMP-2 entrapped in hyaluronic acid-bisphosphonates hydrogel maintains its bioactivity, as shown by the induction of osteogenic differentiation of mesenchymal stem cells.

A similar approach toward bone tissue regeneration was proposed by Furukawa and co-workers [70], who developed a hyaluronic-based hydrogel functionalised with inorganic polyphosphate moieties. The poly-phosphates groups were able to provide osteoconductive stimulation to murine osteoblast precursor cells, demonstrated by the up-regulation of osteogenic marker genes and increased alkaline phosphatase activity. In addition, it was shown that the bioactivity imparted by immobilised pyrophosphates was higher if compared to free polyphosphates embedded within the gel.

Dexamethasone, a synthetic corticosteroid, was also used as a small-molecule for the stimulation of osteogenic differentiation [71,72]. Thus, PEG-based hydrogels were functionalised with dexamethasone through a Diels-Alder conjugation strategy. The reversible nature of the Diels-Alder reaction was exploited for controlling dexamethasone release from the hydrogel. Both in 2D and in 3D hMSCs cell culture, dexamethasone release promoted a significant increase in alkaline phosphatase activity and mineral deposition if compared to that of control gels without dexamethasone, or with dexamethasone in the free form.

## 5. Outlook and Perspectives

The field of hydrogels started with the pioneering Wichterle and Lim in the 1960s [73,74]. Since then, a remarkable development of hydrogels from simple chemically or physically crosslinked networks to complex bioresponsive systems was observed. Although not cited in this review, hydrogels are gaining a high level of sophistication, reflected, for example, in shape memory and self-healing hydrogels [75,76]. The clinical need for easy administration in regenerative medicine applications fuelled the research of injectable hydrogels and bioresponsive constructs able to drive cell response suitable for minimally invasive treatments [77]. The selection of the cross-linking strategy is driven by the need of an immediate change from a low viscous solution before injection and quick formation of a strong network in situ. In addition, the possibility to modulate the degradation profiles after hydrogel administration and the bioactivation strategy can further improve the clinical translation of these scaffolds for tissue engineering applications. It is expected in the next years that more sophisticated hydrogels suitably tuned to sustain and promote adhesion, migration, and differentiation of specific cell lines will be created, bringing cell therapies in tissue regeneration closer to clinical application.

**Acknowledgments:** We gratefully acknowledge the European Community's programme under Grant Agreement number: 642028-H2020-MSCA-ITN-2014 "NABBA" for financial support.

**Author Contributions:** All the authors gathered the literature material, wrote the review, and assembled the figure panels.

**Conflicts of Interest:** The authors declare no conflict of interest.

## References

1. Wilson, A.N.; Giuseppi-Elie, A. Bioresponsive hydrogels. *Adv. Healthcare Mater.* **2013**, *2*, 520–532. [CrossRef] [PubMed]
2. Wu, Y.-L.; Chen, X.; Wang, W.; Loh, X.J. Engineering bioresponsive hydrogels toward healthcare applications. *Macromol. Chem. Phys.* **2016**, *217*, 175–188. [CrossRef]
3. Ebara, M.; Kotsuchibashi, Y.; Narain, R.; Idota, N.; Kim, Y.-J.; Hoffman, J.M.; Uto, K.; Aoyagi, T.M. *Smart Biomaterials*; Springer: Tokyo, Japan, 2014.
4. Koetting, M.C.; Peters, J.T.; Steichen, S.D.; Peppas, N.A. Stimulus-responsive hydrogels: Theory, modern advances, and applications. *Mater. Sci. Eng. R Rep.* **2015**, *93*, 1–49. [CrossRef] [PubMed]
5. Seliktar, D. Designing cell-compatible hydrogels for biomedical applications. *Science* **2012**, *336*, 1124–1128. [CrossRef] [PubMed]
6. Carlini, A.S.; Adamiak, L.; Gianneschi, N.C. Biosynthetic polymers as functional materials. *Macromolecules* **2016**, *49*, 4379–4394. [CrossRef] [PubMed]
7. Russo, L.; Cipolla, L. Glycomics: New challenges and opportunities in regenerative medicine. *Chem. Eur. J.* **2016**, *22*, 13380–13388. [CrossRef] [PubMed]
8. Grim, J.C.; Marozas, I.A.; Anseth, K.S. Thiol-ene and photo-cleavage chemistry for controlled presentation of biomolecules in hydrogels. *J. Control. Release* **2015**, *219*, 95–106. [CrossRef] [PubMed]
9. Collins, M.N.; Birkinshaw, C. Hyaluronic acid based scaffolds for tissue engineering—A review. *Carbohydr. Polym.* **2013**, *92*, 1262–1279. [CrossRef] [PubMed]
10. Croisier, F.; Jérôme, C. Chitosan-based biomaterials for tissue engineering. *Eur. Polym. J.* **2013**, *49*, 780–792. [CrossRef]
11. Bidarra, S.J.; Barrias, C.C.; Granja, P.L. Injectable alginate hydrogels for cell delivery in tissue engineering. *Acta Biomater.* **2014**, *10*, 1646–1662. [CrossRef] [PubMed]
12. Burdick, J.A.; Prestwich, G.D. Hyaluronic acid hydrogels for biomedical applications. *Adv. Mater.* **2011**, *23*, H41–H56. [CrossRef] [PubMed]
13. Zhu, J.; Marchant, R.E. Design properties of hydrogel tissue-engineering scaffolds. *Expert Rev. Med. Devices* **2011**, *8*, 607–626. [CrossRef] [PubMed]
14. Ahmadi, F.; Oveisi, Z.; Samani, S.M.; Amoozgar, Z. Chitosan based hydrogels: Characteristics and pharmaceutical applications. *Res. Pharm. Sci.* **2015**, *10*, 1–16. [PubMed]

15. Lee, K.Y.; Mooney, D.J. Alginate: Properties and biomedical applications. *Prog. Polym. Sci.* **2012**, *37*, 106–126. [CrossRef] [PubMed]

16. Jonker, A.M.; Löwik, D.W.P.M.; van Hest, J.C.M. Peptide- and protein-based hydrogels. *Chem. Mater.* **2012**, *24*, 759–773. [CrossRef]

17. Rafat, M.; Koh, L.B.; Islam, M.M.; Liedberg, B.O.; Griffith, M. Highly elastic epoxy cross-linked collagen hydrogels for corneal tissue engineering. *Acta Ophtalmol.* **2012**, *90*, s249. [CrossRef]

18. Mu, C.; Zhang, K.; Lin, W.; Li, D. Ring-opening polymerization of genipin and its long-range crosslinking effect on collagen hydrogel. *J. Biomed. Mater. Res. A* **2013**, *101*, 385–393. [CrossRef] [PubMed]

19. Zhao, X.; Lang, Q.; Yildirimer, L.; Lin, Z.Y.; Cui, W.; Annabi, N.; Ng, K.W.; Dokmeci, M.R.; Ghaemmaghami, A.M.; Khademhosseini, A. Photocrosslinkable gelatin hydrogel for epidermal tissue engineering. *Adv. Healtcare Mater.* **2016**, *5*, 108–118. [CrossRef] [PubMed]

20. Bajaj, P.; Schweller, R.M.; Khademhosseini, A.; West, J.L.; Bashi, R. 3D Biofabrication strategies for tissue engineering and regenerative medicine. *Annu. Rev. Biomed. Eng.* **2014**, *16*, 247–276. [CrossRef] [PubMed]

21. Greene, T.; Lin, C.-C. Modular cross-linking of gelatin-based thiol-norbornene hydrogels for in vitro 3D culture of hepatocellular carcinoma cells. *ACS Biomater. Sci. Eng.* **2015**, *1*, 1314–1323. [CrossRef]

22. Russo, L.; Sgambato, A.; Visone, R.; Occhetta, P.; Moretti, M.; Rasponi, M.; Nicotra, F.; Cipolla, L. Gelatin hydrogels via thiol-ene chemistry. *Monatsh. Chem.* **2016**, *147*, 587–592. [CrossRef]

23. Xu, K.; Cantu, D.A.; Fu, Y.; Kim, J.; Zheng, X.; Hematti, P.; Kao, W.J. Thiol-ene Michael-type formation of gelatin/poly(ethylene glycol) biomatrices for three-dimensional mesenchymal stromal/stem cell administration to cutaneous wounds. *Acta Biomater.* **2013**, *9*, 8802–8814. [CrossRef] [PubMed]

24. Tian, H.; Tang, Z.; Zhuang, X.; Chen, X.; Jing, X. Biodegradable synthetic polymers: Preparation, functionalization and biomedical application. *Prog. Polym. Sci.* **2012**, *37*, 237–280. [CrossRef]

25. Hunt, J.A.; Chen, R.; van Veen, T.; Bryan, N. Hydrogels for tissue engineering and regenerative medicine. *J. Mater. Chem. B* **2014**, *2*, 5319–5338. [CrossRef]

26. Lee, J.; Bae, Y.H.; Sohn, Y.S.; Jeong, B. Thermogelling aqueous solutions of alternating multiblock copolymers of poly(L-lactic acid) and poly(ethylene glycol). *Biomacromolecules* **2006**, *7*, 1729–1734. [CrossRef] [PubMed]

27. Bellis, S.L. Advantages of RGD peptides for directing cell association with biomaterials. *Biomaterials* **2011**, *32*, 4205–4210. [CrossRef] [PubMed]

28. Ventre, M.; Netti, P.A. Controlling cell functions and fate with surfaces and hydrogels: The role of material features in cell adhesion and signal transduction. *Gels* **2016**, *2*, 12. [CrossRef]

29. Kyburz, K.A.; Anseth, K.S. Synthetic mimics of the extracellular matrix: How simple is complex enough? *Ann. Biomed. Eng.* **2015**, *43*, 489–500. [CrossRef] [PubMed]

30. Zhang, H.; Dai, S.; Bi, J.; Liu, K.-K. Biomimetic three-dimensional microenvironment for controlling stem cell fate. *Interface Focus* **2011**, *1*, 792–803. [CrossRef] [PubMed]

31. Swinehart, I.T.; Badylak, S.F. Extracellular matrix bioscaffolds in tissue remodeling and morphogenesis. *Dev. Dyn.* **2016**, *245*, 351–360. [CrossRef] [PubMed]

32. Tong, X.; Yang, F. Engineering interpenetrating network hydrogels as biomimetic cell niche with independently tunable biochemical and mechanical properties. *Biomaterials* **2014**, *35*, 1807–1815. [CrossRef] [PubMed]

33. Ahadian, S.; Sadeghian, R.B.; Salehi, S.; Ostrovidov, S.; Bae, H.; Ramalingam, M.; Khademhosseini, A. Bioconjugated hydrogels for tissue engineering and regenerative medicine. *Bioconjug. Chem.* **2015**, *26*, 1984–2001. [CrossRef] [PubMed]

34. Jabbari, E. Bioconjugation of hydrogels for tissue engineering. *Curr. Opin. Biotechnol.* **2011**, *22*, 655–660. [CrossRef] [PubMed]

35. Russo, L.; Sgambato, A.; Lecchi, M.; Pastori, V.; Raspanti, M.; Natalello, A.; Doglia, S.M.; Nicotra, F.; Cipolla, L. Neoglucosylated collagen matrices drive neuronal cells to differentiate. *ACS Chem. Neurosci.* **2014**, *5*, 261–265. [CrossRef] [PubMed]

36. Sgambato, A.; Russo, L.; Montesi, M.; Panseri, S.; Marcacci, M.; Caravà, E.; Raspanti, M.; Cipolla, L. Different sialoside epitopes on collagen film surfaces direct mesenchymal stem cell fate. *ACS Appl. Mater. Interfaces* **2016**, *8*, 14952–14957. [CrossRef] [PubMed]

37. Cipolla, L.; Russo, L.; Shaikh, N.; Nicotra, F. Chapter 27: Materials biofunctionalization for tissue regeneration. In *Polymeric Biomaterials*, 3rd ed.; CRC Press: Boca Raton, FL, USA, 2013; Volume 2, pp. 715–736, ISBN: 9781420094725.

38. Petka, W.A.; Harden, J.L.; McGrath, K.P.; Wirtz, D.; Tirrell, D.A. Reversible hydrogels from self-assembling artificial proteins. *Science* **1998**, *281*, 389–392. [CrossRef] [PubMed]

39. Vermonden, T.; Censi, R.; Hennink, W.E. Hydrogels for protein delivery. *Chem. Rev.* **2012**, *112*, 2853–2888. [CrossRef] [PubMed]

40. Censi, R.; Di Martino, P.; Vermonden, T.; Hennink, D.W. Hydrogels for protein delivery in tissue engineering. *J. Control. Release* **2012**, *161*, 680–692. [CrossRef] [PubMed]

41. Lin, C.C.; Anseth, K.S. Cell–cell communication mimicry with poly(ethylene glycol) hydrogels for enhancing β-cell function. *Proc. Natl. Acad. Sci. USA* **2011**, *108*, 6380–6385. [CrossRef] [PubMed]

42. Kim, J.; Lin, B.; Kim, S.; Choi, B.; Evseenko, D.; Lee, M. TGF-β1 conjugated chitosan collagen hydrogels induce chondrogenic differentiation of human synovium-derived stem cells. *J. Biol. Eng.* **2015**, *9*, 1. [CrossRef] [PubMed]

43. Jha, A.K.; Tharp, K.M.; Ye, J.; Santiago-Ortiz, J.L.; Jackson, W.M.; Stahl, A.; Schaffer, D.V.; Yeghiazarians, Y.; Healy, K.E. Enhanced survival and engraftment of transplanted stem cells using growth factor sequestering hydrogels. *Biomaterials* **2015**, *47*, 1–12. [CrossRef] [PubMed]

44. Prokoph, S.; Chavakis, E.; Levental, K.R.; Zieris, A.; Freudenberg, U.; Dimmeler, S.; Werner, C. Sustained delivery of SDF-1α from heparin-based hydrogels to attract circulating pro-angiogenic cells. *Biomaterials* **2012**, *33*, 4792–4800. [CrossRef] [PubMed]

45. Rabbany, S.Y.; Pastore, J.; Yamamoto, M.; Miller, T.; Rafii, S.; Aras, R.; Penn, M. Continuous delivery of stromal cell-derived factor-1 from alginate scaffolds accelerates wound healing. *Cell Transplant.* **2010**, *19*, 399–408. [CrossRef] [PubMed]

46. Wang, T.Y.; Forsythe, J.S.; Parish, C.L.; Nisbet, D.R. Biofunctionalisation of polymeric scaffolds for neural tissue engineering. *J. Biomater. Appl.* **2012**, *27*, 369–390. [CrossRef] [PubMed]

47. Tam, R.Y.; Fuehrmann, T.; Mitrousis, N.; Shoichet, M.S. Regenerative therapies for central nervous system diseases: A biomaterials approach. *Neuropsychopharmacol. Rev.* **2014**, *39*, 169–188. [CrossRef] [PubMed]

48. Mosiewicz, K.A.; Kolb, L.; van der Vlies, A.J.; Martino, M.M.; Lienemann, P.S.; Hubbell, J.A.; Ehrbar, M.; Lutolf, M.P. In situ cell manipulation through enzymatic hydrogel photopatterning. *Nat. Mater.* **2013**, *12*, 1072–1078. [CrossRef] [PubMed]

49. Wheeldon, I.; Farhadi, A.; Bick, A.G.; Jabbari, E.; Khademhosseini, A. Nanoscale tissue engineering: Spatial control over cell-materials interactions. *Nanotechnology* **2011**, *22*, 212001–212017. [CrossRef] [PubMed]

50. Hersel, U.; Dahmen, C.; Kessler, H. RGD modified polymers: Biomaterials for stimulated cell adhesion and beyond. *Biomaterials* **2003**, *24*, 4385–4415. [CrossRef]

51. Zhu, J. Bioactive modification of poly(ethylene glycol) hydrogels for tissue engineering. *Biomaterials* **2010**, *31*, 4639–4656. [CrossRef] [PubMed]

52. Mehta, M.; Madl, C.M.; Lee, S.; Duda, G.N.; Mooney, D.J. The collagen I mimetic peptide DGEA enhances an osteogenic phenotype in mesenchymal stem cells when presented from cell-encapsulating hydrogels. *J. Biomed. Mater. Res. A* **2015**, *103*, 3516–3525. [CrossRef] [PubMed]

53. DeForest, C.A.; Anseth, K.S. Cytocompatible click-based hydrogels with dynamically-tunable properties through orthogonal photoconjugation and photocleavage reactions. *Nat. Chem.* **2011**, *3*, 925–931. [CrossRef] [PubMed]

54. Lee, J.W.; Park, Y.J.; Lee, S.J.; Lee, S.K.; Lee, K.Y. The effect of spacer arm length of an adhesion ligand coupled to an alginate gel on the control of fibroblast phenotype. *Biomaterials* **2010**, *31*, 5545–5551. [CrossRef] [PubMed]

55. Li, J.; Shu, Y.; Hao, T.; Wang, Y.; Qian, Y.; Duan, C.; Sun, H.; Lin, Q.; Wang, C. A chitosan-glutathione based injectable hydrogel for suppression of oxidative stress damage in cardiomyocytes. *Biomaterials* **2013**, *34*, 9071–9081. [CrossRef] [PubMed]

56. Miklas, J.W.; Dallabrida, S.M.; Reis, L.A.; Ismail, N.; Rupnick, M.; Radisic, M. QHREDGS enhances tube formation, metabolism and survival of endothelial cells in collagen-chitosan hydrogels. *PLoS ONE* **2013**, *8*, e72956. [CrossRef] [PubMed]

57. Madl, C.M.; Mehta, M.; Duda, G.N.; Heilshorn, S.C.; Mooney, D.J. Presentation of BMP-2 mimicking peptides in 3D hydrogels directs cell fate commitment in osteoblasts and mesenchymal stem cells. *Biomacromolecules* **2014**, *15*, 445–455. [CrossRef] [PubMed]

58. Dhoot, N.O.; Tobias, C.A.; Fischer, I.; Wheatley, M.A. Peptide-modified alginate surfaces as a growth permissive substrate for neurite outgrowth. *J. Biomed. Mater. Res. A* **2004**, *71*, 191–200. [CrossRef] [PubMed]

59. Seif-Naraghi, S.B.; Horn, D.; Schup-Magoffin, P.J.; Christman, K.L. Injectable extracellular matrix derived hydrogel provides a platform for enhanced retention and delivery of a heparin-binding growth factor. *Acta Biomater.* **2012**, *8*, 3695–3703. [CrossRef] [PubMed]

60. Foster, E.; You, J.; Siltanen, C.; Patel, D.; Haque, A.; Anderson, L.; Revzin, A. Heparin hydrogel sandwich cultures of primary hepatocytes. *Eur. Polym. J.* **2015**, *72*, 726–735. [CrossRef]

61. Li, Z.; Qu, T.; Ding, C.; Ma, C.; Sun, H.; Li, S.; Liu, X. Injectable gelatin derivative hydrogels with sustained vascular endothelial growth factor release for induced angiogenesis. *Acta Biomater.* **2015**, *13*, 88–100. [CrossRef] [PubMed]

62. Lyssiotis, C.A.; Lairson, L.L.; Boitano, A.E.; Wurdak, H.; Zhu, S.; Schultz, P.G. Chemical control of stem cell fate and developmental potential. *Angew. Chem. Int. Ed.* **2011**, *50*, 200–242. [CrossRef] [PubMed]

63. Park, K.M.; Shin, Y.M.; Joung, Y.K.; Shin, H.; Park, K.D. In situ forming hydrogels based on tyramine conjugated 4-Arm-PPO-PEO via enzymatic oxidative reaction. *Biomacromolecules* **2010**, *11*, 706–712. [CrossRef] [PubMed]

64. Prodanovic, O.; Spasojevic, D.; Prokopijevic, M.; Radotic, K.; Markovic, N.; Blazic, M.; Prodanovic, R. Tyramine modified alginates via periodate oxidation for peroxidase induced hydrogel formation and immobilization. *React. Funct. Polym.* **2015**, *93*, 77–83. [CrossRef]

65. Hou, J.; Li, C.; Guan, Y.; Zhang, Y.; Zhu, X.X. Enzymatically crosslinked alginate hydrogels with improved adhesion properties. *Polym. Chem.* **2015**, *6*, 2204–2213. [CrossRef]

66. Yavvari, P.S.; Srivastava, A. Robust, self-healing hydrogels synthesised from catechol rich polymers. *J. Mater. Chem. B* **2015**, *3*, 899–910. [CrossRef]

67. Ryu, J.H.; Lee, Y.; Kong, W.H.; Kim, T.G.; Park, T.G.; Lee, H. Catechol-functionalized chitosan/pluronic hydrogels for tissue Adhesives and hemostatic materials. *Biomacromolecules* **2011**, *12*, 2653–2659. [CrossRef] [PubMed]

68. Zhang, Y.; Kuang, Y.; Gao, Y.; Xu, B. Versatile small molecule motifs for self-assembly in water and formation of bifunctional supramolecular hydrogels. *Langmuir* **2011**, *27*, 529–537. [CrossRef] [PubMed]

69. Hulsart-Billström, G.; Yuen, P.K.; Marsell, R.; Hilborn, J.; Larsson, S.; Ossipov, D. Bisphosphonate-linked hyaluronic acid hydrogel sequesters and enzymatically releases active bone morphogenetic protein-2 for Induction of osteogenic differentiation. *Biomacromolecules* **2013**, *14*, 3055–3063. [CrossRef] [PubMed]

70. Wu, A.T.H.; Aoki, T.; Sakoda, M.; Ohta, S.; Ichimura, S.; Ito, T.; Ushida, T.; Furukawa, K.S. Enhancing osteogenic differentiation of MC3T3-E1 cells by immobilizing inorganic polyphosphate onto hyaluronic acid hydrogel. *Biomacromolecules* **2015**, *16*, 166–173. [CrossRef] [PubMed]

71. Koehler, K.C.; Alge, D.L.; Anseth, K.S.; Bowman, C.N. A Diels-Alder modulated approach to control and sustain the release of dexamethasone and induce osteogenic differentiation of human mesenchymal stem cells. *Biomaterials* **2013**, *34*, 4150–4158. [CrossRef] [PubMed]

72. Yang, C.; Mariner, P.D.; Nahreini, J.N.; Anseth, K.S. Cell-mediated delivery of glucocorticoids from thiol-ene hydrogels. *J. Control. Release* **2012**, *162*, 612–618. [CrossRef] [PubMed]

73. Wichterle, O.; Lim, D. Hydrophilic gels for biological use. *Nature* **1960**, *185*, 117–118. [CrossRef]

74. Buwalda, S.J.; Boere, K.W.; Dijkstra, P.J.; Feijen, J.; Vermonden, T.; Hennink, W.E. Hydrogels in a historical perspective: From simple networks to smart materials. *J. Control. Release* **2014**, *190*, 254–273. [CrossRef] [PubMed]

75. Miyamae, K.; Nakahata, M.; Takashima, Y.; Harada, A. Self-healing, expansion-contraction, and shape-memory properties of a preorganized supramolecular hydrogel through host-guest interactions. *Angew. Chem. Int. Ed.* **2015**, *54*, 8984–8987. [CrossRef] [PubMed]

76. Gulyuz, U.; Okay, O. Self-healing poly(acrylic acid) hydrogels with shape memory behaviour of high mechanical strength. *Macromolecules* **2014**, *47*, 6889–6899. [CrossRef]

77. Toh, W.S.; Loh, X.J. Advances in hydrogel delivery systems for tissue regeneration. *Mater. Sci. Eng.* **2014**, *45*, 690–697. [CrossRef] [PubMed]

*gels*

MDPI

*Review*

# Hydrogels for Biomedical Applications: Their Characteristics and the Mechanisms behind Them

Qinyuan Chai, Yang Jiao and Xinjun Yu *

Department of Chemistry, University of Cincinnati, Cincinnati, OH 45221, USA; chaiqn@mail.uc.edu (Q.C.); jiaoyn@mail.uc.edu (Y.J.)
* Correspondence: yux3@mail.uc.edu; Tel.: +1-513-556-1107

Academic Editor: David Díaz Díaz
Received: 16 August 2016; Accepted: 11 January 2017; Published: 24 January 2017

**Abstract:** Hydrogels are hydrophilic, three-dimensional networks that are able to absorb large quantities of water or biological fluids, and thus have the potential to be used as prime candidates for biosensors, drug delivery vectors, and carriers or matrices for cells in tissue engineering. In this critical review article, advantages of the hydrogels that overcome the limitations from other types of biomaterials will be discussed. Hydrogels, depending on their chemical composition, are responsive to various stimuli including heating, pH, light, and chemicals. Two swelling mechanisms will be discussed to give a detailed understanding of how the structure parameters affect swelling properties, followed by the gelation mechanism and mesh size calculation. Hydrogels prepared from natural materials such as polysaccharides and polypeptides, along with different types of synthetic hydrogels from the recent reported literature, will be discussed in detail. Finally, attention will be given to biomedical applications of different kinds of hydrogels including cell culture, self-healing, and drug delivery.

**Keywords:** hydrogels; biomaterials; drug delivery; self-healing; cell culture

## 1. Introduction

Hydrogels, crosslinked 3D networks of hydrophilic polymer chains, are capable of holding large amounts of water due to their hydrophilic structure [1–4]. Thus, the hydrogel networks can extensively swell in water media. Since water is the greatest component of the human body, a hydrogel, which can absorb large quantities of water, is considered to have great potential when applied for biomedical purposes [5–10]. Recently, wide investigation has been going on into the feasibility of applying hydrogels in fields including tissue engineering, drug delivery, self-healing materials, biosensors, and hemostasis bandages [11–15]. Compared with other types of biomaterials, hydrogels have the advantages of increased biocompatibility, tunable biodegradability, properly mechanical strength, porous structure, and so on. However, due to the low mechanical strength and fragile nature of the hydrogels, the feasibility of applying hydrogels is still limited. Thus, novel hydrogels with stronger and more stable properties are still needed and remain an important direction for research.

As expected, naturally formed hydrogels are gradually replaced by synthetic hydrogels to achieve longer service life, high capacity of water absorption, and high gel strength [8,9]. Fortunately, with various developed synthetic strategies, hydrogels with defined network structures, desirable chemical compositions, and tunable mechanical strength can be designed. Hydrogels can be prepared from completely artificial components and show remarkable stability even under severe conditions such as high temperature or a very acidic or basic environment. Additionally, by modifying the polymer chains with stimuli-responsive functional groups, the hydrogel properties can be switched by stimuli including heat, light, magnetic fields, chemical agents, and pH [16–20].

In this critical review, different technologies for preparing hydrogels will be discussed and we will take a closer look at different kinds of stimuli-responsive hydrogels. Detailed swelling mechanisms proposed based on various theories will be covered to give a deeper understanding of such materials. Last but not least, attention will be given to different hydrogels to achieve specific requirements for biomedical purposes such as cell culture, self-healing, and drug delivery.

## 2. Different Kinds of Stimuli-Responsive Hydrogels

Defined by Peppas [21], "hydrogels are hydrophilic, three-dimension networks, which are able to imbibe large amounts of water or biological fluids, and thus resemble, to a large extent, a biological tissue". They are insoluble in any solvent due to the polymer chains being crosslinked by either covalent bonds or physical interactions such as entanglements and crystallites [22–24]. Due to the properties of hydrogels, such as high content of water, soft and rubbery consistence, as well as low interfacial tension with water or biological fluids, they are expected to be potential alternatives for natural tissues [25]. According to different applications, the hydrogel can be prepared to respond to various stimuli in the body such as pH, ionic strength, and temperature.

### 2.1. Thermoresponsive Hydrogels

The equilibrium between the hydrophobic and hydrophilic segments is the key to control the properties of a synthetic thermoresponsive hydrogel. In detail, the temperature has a remarkable effect on the hydrophobic interactions between hydrophobic polymer segments and the hydrophilic interactions between hydrophilic polymer segments and water molecules. Thus, a small temperature change can interrupt the original equilibrium and induce sol–gel transition [26]. In the work done by Vernon's group [27], they synthesized a temperature-responsive graft copolymer based on *N*-isopropylacrylamide (NIPAAm) and Jeffamine M-1000 acrylamide (JAAm), which showed controlled swelling properties without introducing degradable moieties or increasing the lower critical solution temperature (LCST) above body temperature. The JAAm was hypothesized to be useful as a component in the polymer architecture to control the swelling and release properties with minimal effect on LCST. The outstanding hydrolytic stability, hydrophilicity, and minimal LCST effect make this hydrogel a suitable candidate for a variety of temperature-responsive biomaterials where control over swelling or drug release is crucial.

In another work [28], Long and co-workers investigated the application of a poly(*N*-isopropylacrylmide) (PNIPAAm) hydrogel thin film to thermochromic smart windows. The solar modulating ability ($\Delta T_{sol}$) showed an ultra-large value with the luminous transmittance ($T_{lum}$) to be highest identified so far. This hydrogel demonstrated an excellent combination of high $T_{lum}$ along with dramatically improved solar modulating ability, which may lead to the development of thermochromic smart windows based on organic materials.

### 2.2. pH-Responsive Hydrogels

pH-Responsive hydrogels are a class of biomaterials that exhibit desirable physical and chemical properties at specific pH ranges. Acidic or basic groups are bonded to the polymer chains. The acidic groups deprotonate at high pH, while the basic groups protonate at low pH. The association, dissociation, and binding of various ions to polymer chains cause hydrogel swelling in an aqueous solution.

Peppa's group fabricated a hydrophilic pH-responsive hydrogel based on poly(methyacrylic-*graft*-ethylene glycol) (P(MMA-*g*-EG)) conjugated with hydrophobic PMMA nanoparticles [29]. By incorporating a different mole ratio of PMMA nanoparticles in the P(MMA-*g*-EG), it forms amphiphilic polymer carriers with tunable physical properties. The release of encapsulated therapeutic agents triggered by a change of pH from the stomach to the small intestine was tested. Furthermore, the cyto-compatibility of the polymer materials was investigated on cells modeling the gastrointestinal

(GI) tract and colon cancer cells. This pH-responsive nanoparticle containing P(MMA-*g*-EG) provides the possibility to be used as oral delivery vectors of chemotherapeutics for cancer.

In general, the size of the gel will respond to environment pH as well as salt concentration. Thus, an equilibrium model was established by Moore's group to predict the swelling/deswelling behavior of hydrogels in different pH solutions. The validation of the model was conducted by comparing the simulations with experimental results. This model was then utilized to investigate the effects of different hydrogels and solution conditions on the degree and rate of swelling/deswelling of those hydrogels [30]. It was found that the higher the concentration and buffer diffusivity are, the faster the kinetic. All these parameters can be used to tune the performance of hydrogel microactuators, suggesting that the mechanical properties of the hydrogel can be varied considerably by varying the pH of solutions.

## 2.3. Light- and Chemical-Responsive Hydrogels

Light-responsive hydrogels are promising functional materials for potential application in the areas of drug/gene delivery [31], micro lenses [32], sensors [33], etc. due to the fact that the activation process via light can be remote and noninvasive. The prepared hydrogel consists of a deoxycholic acid-modified β-cyclodextrin derivative and an azobenzen-branched poly(acrylic acid) copolymer, and can be converted efficiently from gel to sol phase upon photo irradiation with light of 355 nm. The hydrogel was able to recover from sol to gel phase upon photo irradiation with light of 450 nm (Figure 1) [34]. The reversible transition of this hydrogel can be controlled under mild condition, suggesting that this gel material has a promising role in bioengineering applications for the release of molecular and cellular species.

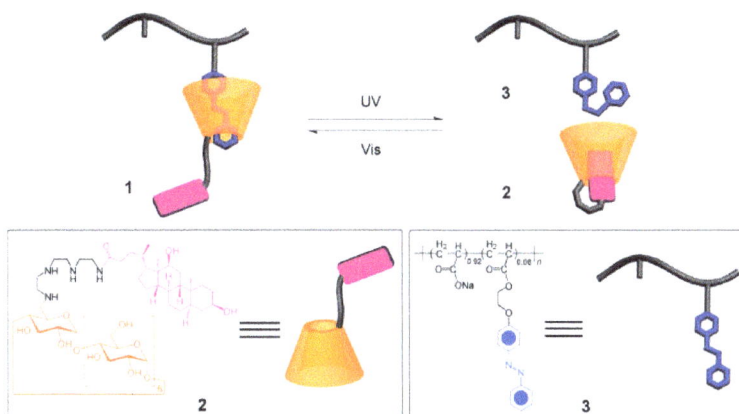

**Figure 1.** Supramolecular inclusion complex **1** formed from deoxycholate-β-CD derivative **2** and azobenzene-branched poly(acrylic acid) copolymer **3**. Reprinted from [34] with permission from the American Chemical Society (2009).

In another example, a novel light-responsive hydrogel was made from a poly(*N*-isopropylacrylamide) (PNIPAAm) nanocomposite incorporating glycidyl methacrylate functionalized graphene oxide (GO-GMA). As a result, the nanocomposite hydrogel will undergo a large volume change under stimulation of infrared (IR) light, because of the highly efficient photo thermal conversion of GO-GMA. This material has the potential to be used as an actuator in microelectromechanical systems or microfluidic devices [35].

In a different work done by Maeda's group [36], they reported DNA-responsive hydrogels that "only shrunk" by the addition of ssDNA (single-stranded DNA) samples (Figure 2).

This biomaterial was developed by polyacrylamide (polyAAm) hydrogels containing directly grafted ssDNA or an ssDNA-polyAAm conjugate in a semi-interpenetrating network. Unlike traditional stimuli-responsive hydrogels, this conjugate retains the advantage of using cross-linkable ssDNAs with well-characterized conformational properties, thus providing potential applications in DNA sensing or DNA-triggered actuators.

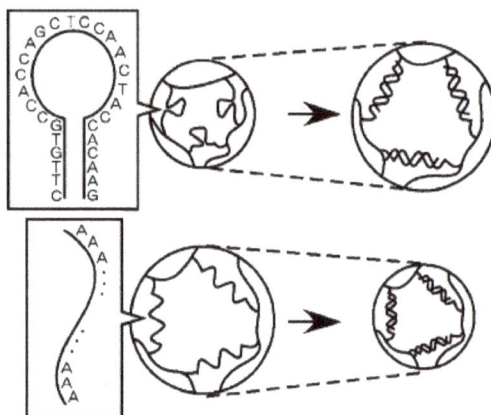

**Figure 2.** The response of novel hybrid hydrogels containing ssDNA as a cross-linker to ssDNA. Reprinted from [36] with permission from the American Chemical Society (2005).

### 3. Different Theories behind the Hydrogel Swelling Mechanism

The properties of hydrogels for specific applications depend on their bulk structures. For network structure characterizations, there are several important parameters such as volume fraction in the swollen state, the corresponding mesh size, and the molecular weight of the polymer chain between neighboring crosslink points. The volume fraction of polymers in the swollen state is a parameter describing how much fluid can be absorbed and retained. The molecular weight between neighboring crosslink points, either covalent bond or physical interaction, is a parameter describing the degree of crosslinking. These parameters are related to each other and can be calculated theoretically or determined by a variety of experimental techniques. In the following paragraph, the two most widely used methods, the equilibrium swelling theory and the rubber elasticity theory, will be discussed.

*3.1. Equilibrium Swelling Theory*

The Flory–Rehner equation describes the mixing of polymers and liquid molecules, which can be used to analyze hydrogels without ionic domains [37]. The equilibrium status of the hydrogel swollen in a fluid is determined by two reverse forces. One is the thermodynmical force of mixing favors swelling, while the other is the stored force in the stretched polymer chains hindering swelling [30].

These two forces balance out each other as described in Equation (1) for the physical situation in terms of the Gibbs free energy:

$$\Delta G_{total} = \Delta G_{elastic} + \Delta G_{mixing} \tag{1}$$

where $\Delta G_{elastic}$ comes from the elastic stored forces in the extended polymer chains contained in the gel networks; $\Delta G_{mixing}$ is the result from the mixing between fluid molecules with the polymer chains. The mixing factor is a measure of the compatibility of the polymer with the solvent molecules, which is usually expressed by the polymer–solvent interaction parameter, $\chi$ [38].

Differentiation of Equation (1) with respect to the number of solvent molecules, while keeping the temperature and pressure constant, gives Equation (2):

$$\mu_1 - \mu_{1,o} = \Delta\mu_{\text{elastic}} + \Delta\mu_{\text{mixing}}. \tag{2}$$

In the equilibrium status, the chemical potential outside the gel should be equal to the chemical potential inside the gel ($\Delta\mu_{1,o} = \Delta\mu_1$). As a result, the chemical potential change from free energy of mixing and elastic force stored in the stretched polymer chains have to cancel out each other.

The previous Flory–Rehner theory was modified for a hydrogel synthesized from aqueous phase. The water contained sufficiently changed chemical potential due to elastic forces, which is responsible for the change of volume fraction density of the polymer chains in the crosslinking process [39]. The presence of ionic moieties in hydrogel makes the situation much more complex, due to thermo complex system from the ionic domain of the polymer chains, which introduces an extra changing factor into the Gibbs free energy.

*3.2. Rubber Elasticity Theory*

From a mechanical perspective, hydrogels assemble natural rubbers that deform elastically in response to applied stress. Treloar [40] and Flory [41] utilized the elastic properties of the hydrogels to describe their structure. However, the original elasticity theory does not apply to hydrogels prepared in solvent. The theory of rubber elasticity by Peppas as in Equation (3) [42] is the only form used to analyze the hydrogel structure, with hydrogels prepared in solvent:

$$\tau = \frac{\varrho RT}{\overline{M_c}} \left(1 - \frac{2\overline{M_c}}{\overline{M_n}}\right) \left(\alpha - \frac{1}{\alpha^2}\right) \left(\frac{v_{2,s}}{v_{2,r}}\right)^{1/3}, \tag{3}$$

where $\tau$ is the applied stress to the polymer sample, $\varrho$ is the density of the polymer, $R$ is the universal gas constant, $T$ is the absolute experimental temperature, and $M_c$ is the molecular weight between crosslinks. To utilize this elasticity theory to analyze the structure of the hydrogel, experiments must be done in the tensile mode [43,44].

*3.3. Mechanism of Gelation*

In the thermally induced sol–gel transition, there are different processes involved including hydrophobic and hydrophilic interactions, coil to helix transition, micelle packing, and so on. To understand the exact gelation mechanism behind certain polymers, figuring out the exact molecular level process is essential [45]. The most widely reported thermally induced gelation is based on the equilibrium between hydrophobic and hydrophilic interactions. For example, introducing a hydrophobic segment such as methyl, ethyl, or propyl to hydrophilic polymers is an efficient way to tune the hydrophobicity of the polymer [46]. LCST is a critical temperature below which the system will be miscible and above which phase separation will occur, forming gels. Interactions between polymer and polymer, polymer and water, and water and water take place in aqueous polymer solutions. The LCST of the system depends on the equilibrium status of these interactions. The most efficient way to determine LCST is by light scattering, with the collapse and aggregation of the polymer chains during gelation state inducing a dramatic increase of the light scattering [47].

Thermodynamically, the thermally induced abrupt change in the solubility is controlled by the Gibbs free energy of mixing [48]. A small change in temperature can cause negative change in the Gibbs free energy. As a result, the interaction between polymer and water will be eliminated and the water–water and polymer–polymer interaction will be favored. To equilibrate this negative Gibbs free energy change, there must be an increase in the entropy term due to the already known increased enthalpy term. Due to the dramatic increase in the hydrophobic interactions between polymer chains, at the sol–gel transition temperature, the polymer chains quickly dehydrate and collapse to a more hydrophobic structure [45,49]. On the other hand, some amphiphilic block copolymers will self-assemble into micelle structures due to the hydrophobic interaction to equilibrate the decrease of the Gibbs free energy [50].

Depending on the concentration, amphiphilic block copolymers can form micelles, which are aggregates of surfactant molecules dispersed in a liquid colloid and hydrogels by adding water and adjusting the temperature. These block copolymers build up a structure with a hydrophobic core and hydrophilic shell with typical micelle size between 20 and 100 nm. All the gelation mechanisms discussed here are based on reversible physical linkage, so the gelation transition is reversible after removing the gelling stimuli.

### 3.4. Calculation of the Mesh Size

The space contained in a hydrogel, responsible for the diffusion properties, is often regarded as the 'pore'. Depending on the size of these pores, hydrogels are commonly classified as macro-pores, micro-porous, or non-porous. The size of the pore is often described by a structural parameter, the correlation length $\xi$, which is defined as the linear distance between two neighboring crosslinks [51]:

$$\xi = \alpha \left( \overline{r}_0^2 \right)^{1/2}. \tag{4}$$

Here, $\alpha$ is the elongation ratio of the polymer chains and $\overline{r}_0$ stands for the distance between two adjacent crosslinking points of the unperturbed polymer chain [52]. From the volume fraction of the swollen polymer $v_{2,s}$, the elongation ratio $\alpha$ can be calculated as:

$$\alpha = v_{2,s}^{-1/3}. \tag{5}$$

The unperturbed end-to-end distance of the polymer chain between two neighboring crosslinks can be calculated by:

$$\left( \overline{r}_0^2 \right)^{1/2} = l(C_n N)^{1/2}, \tag{6}$$

where $l$ is the length of the bond along the polymer backbone (1.54 Å for vinyl polymers), $C_n$ is the Flory characteristic ratio, and $N$, the number of links per chain, can be calculated by:

$$N = \frac{2\overline{M}_c}{M_r}, \tag{7}$$

where $M_r$ is the molecular weight of the repeat unit. Finally, combining all the equations above, the correlated distance of the polymer chains between two adjacent crosslinking points can be evaluated:

$$\xi = v_{2,s}^{-1/3} \left( \frac{2C_n \overline{M}_c}{M_r} \right)^{1/2} l. \tag{8}$$

## 4. Hydrogels Based on Natural Materials

### 4.1. Hydrogels Based on Polysaccharides

Cellulose is a natural polysaccharide that cannot dissolve in water. Unlike other types of water-soluble polysaccharides, cellulose requires separate cross-linking to fabricate a hydrogel network [53]. Native cellulose nanofibers are normally generated from bacterial and plants. Those nanofibers prefer to disperse in an aqueous solution rather than dissolving [54]. In Yliperttula and co-workers' work [55], a plant-derived nanofibrillar cellulose (NFC) hydrogel with desired functionality makes a potential 3D cell culture scaffold. The structural properties of NFC hydrogel were evaluated along with rheological properties, cellular biocompatibility, cellular polarization, and differentiation of human hepatic cell lines. Due to its fluid-like property, NFC was demonstrated to be injectable at high stress, which makes it possible to mix into gels. Furthermore, spontaneous gelation after injection imparted the necessary mechanical support for both cell growth and differentiation.

Another technique was developed to immobilize an enzyme/antibody. After partial oxidation by sodium periodate, a cellulose hydrogel was prepared from an aqueous alkali-urea solvent. This enabled the cellulose gel to further introduce aldehyde groups [56]. By a Schiff base formation between the aldehyde and amino groups of protein, various active proteins can be covalently introduced to a cellulose gel and stabilized by a reduction of imines, which was confirmed by a coloring reaction. The same strategy is applicable to the peroxidase antibody, which makes various active proteins have the ability to be immobilized on cellulose gels by mild and facile processing. Due to the excellent chemical and mechanical stability of cellulose, this strategy and the afforded materials have the potential to be used for biochemical processing and sensing materials.

Reported by Lee and co-workers [57], by the use of a biocompatible ionic liquid, lipase from *Candida rugosa* was successfully trapped into various cellulose–biopolymer composite hydrogels. A biocompatible ionic liquid, 1-ethyl-3-methylimidazolium acetate, was used, which is known to be one of the best solvents for lignocellulosic materials among the ionic liquids (ILs). The lipase was successfully immobilized in various cellulose composite hydrogels, which is the first report that successfully entrapped an enzyme into non-derivatized cellulose–biopolymer composite hydrogels.

### 4.2. Hydrogels Based on Polypeptides

Gelatin is a denatured product of collagen, a mixture of peptides and proteins produced by partial hydrolysis of collagen extracted from the skin, bones, and connective tissues of animals, which is easily available, degradable, and demonstrates good biocompatibility in vivo. Also, gelatin retains cell-binding motifs such as arginylglycylaspartic acid (RGD) and matrix metalloproteinase (MMP)-sensitive degradation sites, which is the critical component in cell encapsulation [58,59].

Kao's group [60] developed an easy strategy using cysteine to modify gelatin through a bifunctional PEG. In this way, free thio groups can be introduced to gelatin chains based on thiolated gelatin and poly(ethylene glycol) diacrylate. Varying concentration and ratio of precursor provides these crosslinked gelatin-based hydrogels with easy mechanical property modulation. In a 3D environment, gelatin crosslinking modality is crucial for long-term integrin binding sites as well as supporting cell attachment and proliferation by cell morphology and proliferation study.

In another study done by Melero-Martin's group [61], they demonstrated that bioengineering human vascular networks inside methacrylated gelatin (GelMA) constructed in a liquid form could be injected into immunodeficient mice, followed by instantaneous crosslinking when exposed to UV light. A solution of GelMA containing human blood-derived endothelial colony-forming cells (ECFCs) and bone marrow-derived mesenchymal stem cells (MSCs) can be injected into the subcutaneous space of an immunodeficient mouse and then rapidly crosslinked with a controllable degree of GelMA through the exposure time to UV light. For future regenerative applications, which require the formation of functional vascular beds in vivo, GelMA is a good way to deliver vascular cells due to its injectable form before crosslinking.

In another example [62], a three-dimensional scaffold containing self-assembled polycaprolactone (PCL) sandwiched in a gelatin-chitosan hydrogel was developed for application as a biodegradable patch for surgical reconstruction of congenital heart defects; it contains a thin, self-assembled PCL core, intended to facilitate handling, cutting, and suturing the material and to provide sufficient tensile strength to function in the ventricular wall. The developed novel hydrogel was demonstrated to have significant potential to be used as a cardiac patch that can repair congenital cardiac defects.

## 5. Synthetic Hydrogels

Synthetic polymer-based hydrogels, due to their widely variable and easily tuned properties, have been extensively studied. By varying the chemical composition and preparation methods, the structure of the hydrogels can be controlled. Beneficial properties including porosity, swelling ability, stability, biocompatibility/biodegradability, and mechanical strength can all be tuned for specific application purposes. For example, vehicles with a controlled release rate for either small molecular or macromolecular drugs including DNA, enzymes, and peptides can be achieved [42].

Filipovic's group [63] synthesized a novel temperature-and pH-sensitive hydrogel based on N-isopropylacrylamide (NIPAAm) and itaconic acid (IA) through free radical polymerization with lipase extracted from *Candida rugosa*, which is a promising system that can be applied as a pH-responsive device for drug delivery purposes. The properties of these prepared hydrogels were found to be highly sensitive to changes of temperature and pH, while keeping the ionic strength constant. A series of hydrogels was prepared with different molar ratios of NIPAAm and IA. Their morphology, mechanical properties, swelling degree, protein loading efficiency, and release rate were all evaluated. The protein release pattern clearly depends on the extent of swelling of the hydrogels.

Vuluga and co-workers reported the synthesis of a novel thermoresponsive crosslinked hydrogel based on different molecular weight of poly(propylene glycol)s (PPG) and diepoxy-terminated poly(ethylene glycol)s (PEG) to control the multi-block copolymer structure [64]. In an ideal situation, the hydrogel structure is expected to contain one PPG block and two PEG chains linked to the same amine group, leading to a structure with each PPG block surrounded by four PEG blocks, while each PEG block has two PPG blocks and two PEG blocks as neighbors. Both thermoresponsive and swelling properties can be adjusted by controlling the molecular weight of the constituent blocks or the salt added in.

For synthetic hydrogels, in addition to single network hydrogels, hydrogels consisting of two independently crosslinked polymer networks attracted wide attention recently, for which tough gels can be formed even with a less crosslinked "second work" within a more highly crosslinked "first network". The molar ratio of the second network repeat units to the first network needs to be >5 [65].

In the work done by Spinks's group [66], a novel double network hydrogel was synthesized with a bottlebrush structure formed from oligo-monomers of poly(ethylene glycol) methyl ether methacrylate as the first polymer network and poly(acrylic acid) as the second network. The strong intermolecular interactions between the neutral poly(ethylene glycol) side chains and the non-ionic groups offer the hydrogel excellent mechanical strength and high sensitivity to pH changes. Such a material with robust nature and sensitivity to pH changes has a potential as artificial muscles or controlled release devices.

In situ gelable interpenetrating double-network hydrogels have been formulated, prepared from thiolated chitosan and oxidized dextran in a one-pot process. No potentially cytotoxic small-molecule cross-linkers are required and they are without complex maneuvers or catalysis. It was reported that the interpenetrating double-network structure enhanced the mechanical properties and gelation performance of the hydrogel formulated from the thiolated chitosan. In conclusion, this hydrogel system is promising as an in situ formable biomaterial for clinically related applications requiring mechanical strength, durability, and fast gelation [67].

A new inorganic–organic double-network hydrogel composing of poly(acrylic acid) and graphene was prepared with the as-prepared 3D graphene architecture to be the first network and acrylic acid monomer dispersed into the consecutive channels and polymerized which is the second network. This inorganic–organic double-network hydrogel shows both flexibility and electrical conductivity, and can be used in the next generation of flexible electric devices [68].

## 6. Applications in Biomedical Field

### 6.1. Hydrogels for Three-Dimensional Cell Culture

Hydrogels, with high water content as well as tissue-like mechanical properties, have been demonstrated to be capable of combining with cells to engineer various tissues in both vitro and vivo [69,70]. A crucial requirement for the construction of three-dimensional regenerative tissue in sufficient quantities is an artificially created environment that enables biological cells to grow or interact with their surroundings in all three dimensions. Anseth's group [71] reported a new cross-linking chemistry by the tetrazine-norbornene click reaction for the formation of cell-laden hydrogels for 3D cell culture (Figure 3). A PEG functionalized with benzylamino tetrazine moiety was specifically chosen and used because in their previous work it was shown to have high reactivity toward norbornene. The bio-orthogonality, ideal reaction kinetics, and amenability to photochemical patterning made this hydrogel platform have potential applications in a variety of fundamental as well as translational tissue engineering applications.

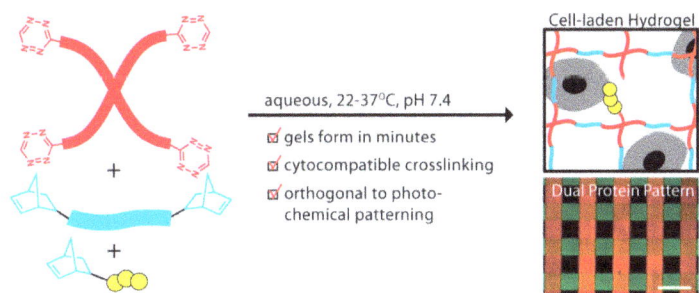

**Figure 3.** Synthetically tractable click hydrogels for three-dimensional cell culture formed using tetrazine–norbornene chemistry. Reprinted from [71] with permission from the American Chemical Society (2013).

Another work was done by Loessner and co-workers [72]. They characterized gelatin methacrylamide (GelMA)-based hydrogels and established them as in vitro and in vivo spheroid-based models for ovarian cancer, to efficiently reflect the advanced disease stage of patients. Hydrogels of equal size, diffusion, and physical properties were generated by employing a controlled preparation and validation protocol. Such GelMA-based hydrogels served as a low-cost, reproducible, and tailorable matrix for 3D cancer cell cultures. Thus, they can be applied as an alternative to improve the understanding of disease progression on a cellular level and also to screen anti-cancer drugs.

### 6.2. Hydrogels for Self-Healing

Self-healing is one of the most outstanding properties of natural materials such as skin, bones, and wood. Thus, hydrogels that can self-heal open up another field for biomedical applications [73,74]. Even though synthetic hydrogels are fabricated to mimic biological tissues, most of the time they still lack the ability to self-heal. This drawback will limit their utilization in many applications that require high stress. As a consequence, researchers dedicate a lot of effort to improving the mechanical properties of hydrogels, including the self-healing property.

The process of healing cracks in natural systems usually contains an energy dissipation mechanism. The self-healing can occur in the presence of sacrificial bonds, which can break and reform dynamically before or during the failure occurring. To prepare a self-healing hydrogel, both covalent [75,76] and non-covalent [77,78] interactions have been reported.

The synthesis of novel hydrogels containing reversible oxime crosslinks were reported by Mukherjee [79]; they are capable of autonomous healing as a consequence of their dynamic nature. To prepare these hydrogels, copolymers containing keto functional groups were synthesized by copolymerizing *N,N*-dimethylacrylamide (DMA) and diacetone acrylamide (DAA) via free radical polymerization. Then the afforded hydrophilic copolymers were covalently crosslinked with difunctional alkoxyamine to obtain hydrogels by the formation of oxime. Along with the efficient self-healing ability, the reversibility of oxime linkages also led to reversible gel-to-sol transitions when excess monofunctional alkoxyamine was added at ambient temperature.

Other than chemical cross-linkers, hydrophobic interactions can also play an important role as a cross-linker for self-healing hydrogels. Okay's group developed a hydrogel by copolymerization of a large hydrophobic monomer stearyl methacrylate and dococyl acrylate with a hydrophilic monomer acrylamide in a micellar solution of sodium dodecyl sulfate [80]. After the addition of salt, micelles grow and solubilize hydrophobes. This hydrogel was demonstrated to have a high degree of toughness due to the finite lifetime of hydrophobic interactions between stearyl methacrylate and dococyl acrylate blocks.

A different work was recently reported by Yamauchi and co-workers [81]. In this research, hydrogels containing cationic substituents were prepared via free-radical polymerization. After applying an aqueous dispersion of Micaromica to the surface, when brought into contact the hydrogels adhered strongly due to the intercalation of cationic substituents included in the gel networks into the interlayers of Micromica. The adhesive strength became higher and was able to support a tensile load of 10 kg as the water content ratio of hydrogels decreased (Figure 4).

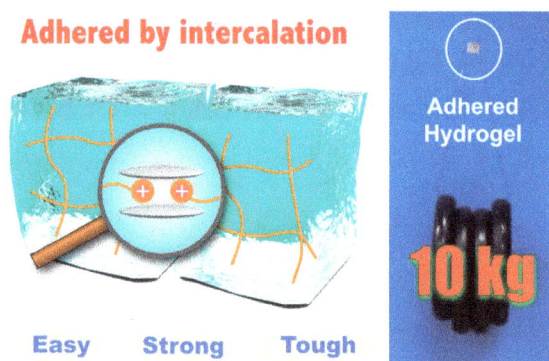

**Figure 4.** Schematic representation of high concentration cationic gels adhered using 1.6 mg of Micromica supporting a tensile load of 10 kg. Reprinted from [81] with permission from the American Chemical Society (2016).

*6.3. Hydrogels for Drug Delivery*

To deliver drugs, porous structure of hydrogels can provide a matrix for drug loading and protect drugs from hostile environment at the same time. Moreover, this porosity can be controlled by varying the crosslinking density of the gel matrix. The release rate, another important parameter for drug carriers, mainly depends on the diffusion coefficient of this molecule through the gel network and can also be tuned according to specific requirements. Biocompatibility and biodegradability can be obtained by designing certain chemical and physical structures for hydrogels. All of those properties lend hydrogels great potential to be used for drug delivery [46,82].

Poly (ethylene oxide)-*b*-poly(propylene oxide)-*b*-poly (ethylene oxide) triblock copolymers (PEO–PPO–PEO) (known as Pluronic or Poloxamer) have been extensively used in pharmaceutical systems [83]. Paavola and co-workers fabricated an injectable gel based on Poloxamer to carry and control the release of the anesthetic agent lidocaine [84]. Since Poloxamer is commercially available, this method was proven to be suitable for hospital utilization. Nevertheless, the relatively rapid diffusion of drugs out of the gel matrix, as well as the duration of drug release, is still limited. It can be improved by covalent crosslinking with other functional group, such as ethoxysilane, amine, or carbohydrates, to prevent the dilution of the polymer in water [85–87].

In addition, to load drug into a gel matrix, conjugating drugs to a hydrogel crosslinked network is another way to deliver drugs. Zhu and co-workers reported a supramolecular based hydrogel to deliver doxorubicin (DOX). First, they synthesized a supramolecular polymeric prodrug via host–guest interaction between cyclodextrin-functionalized polyaldehyde and DOX-modified adamantine. After crosslinking by carboxymethyl chitosan, an injectable DOX-loaded hydrogel was created. This hydrogel was shown to release DOX when exposed to acid stimuli [88].

## 7. Conclusions

Compared with other types of biomaterials, hydrogels have distinct properties such as high water content, controllable swelling behavior, ease of handing, as well as biocompatibility, which makes them attractive for biomedical applications. Based on their chemical structure and crosslink network, hydrogels can respond to different types of stimuli including thermal, pH, light, and chemical stimuli, which can meet various application requirements. Two different hydrogel swelling mechanisms were discussed to give a thorough understanding of how the bulky structure affects the properties of the swollen hydrogels under specific conditions. Hydrogels based on natural materials such as polysaccharides and polypeptides, along with synthetic hydrogels were exampled in detail. Hydrogels, which are three-dimensional crosslinked polymeric networks able to swell in large amounts of water, should be considered prime candidates for carriers or matrices for cells in tissue engineering, self-healing materials, and delivery vehicles for drugs and biomolecules. Further research should be directed to achieving high mechanical strength, fast and efficient self-healing ability, and various biological activities for different biomedical purposes.

**Author Contributions:** Qinyuan Chai, Yang Jiao and Xinjun Yu contributed equally to this work.

**Conflicts of Interest:** The authors declare no conflict of interest.

## References

1. Ahmed, E.M. Hydrogel: Preparation, characterization, and applications: A review. *J. Adv. Res.* **2015**, *6*, 105–121. [CrossRef] [PubMed]
2. Sun, Y.; Kaplan, J.A.; Shieh, A.; Sun, H.-L.; Croce, C.M.; Grinstaff, M.W.; Parquette, J.R. Self-assembly of a 5-fluorouracil-dipeptide hydrogel. *Chem. Commun.* **2016**, *52*, 5254–5257. [CrossRef] [PubMed]
3. Kim, S.H.; Sun, Y.; Kaplan, J.A.; Grinstaff, M.W.; Parquette, J.R. Photo-crosslinking of a self-assembled coumarin-dipeptide hydrogel. *New J. Chem.* **2015**, *39*, 3225–3228. [CrossRef]
4. Verhulsel, M.; Vignes, M.; Descroix, S.; Malaquin, L.; Vignjevic, D.M.; Viovy, J.L. A review of microfabrication and hydrogel engineering for micro-organs on chips. *Biomaterials* **2014**, *35*, 1816–1832. [CrossRef] [PubMed]
5. Daniele, M.A.; Adams, A.A.; Naciri, J.; North, S.H.; Ligler, F.S. Interpenetrating networks based on gelatin methacrylamide and PEG formed using concurrent thiol click chemistries for hydrogel tissue engineering scaffolds. *Biomaterials* **2014**, *35*, 1845–1856. [CrossRef] [PubMed]
6. Yu, X.; Jiao, Y.; Chai, Q. Applications of Gold Nanoparticles in Biosensors. *Nano LIFE* **2016**, *6*, 1642001. [CrossRef]
7. Yu, X.; Chen, X.; Chai, Q.; Ayres, N. Synthesis of polymer organogelators using hydrogen bonding as physical cross-links. *Colloid Polym. Sci.* **2016**, *294*, 59–68. [CrossRef]
8. Billiet, T.; Vandenhaute, M.; Schelfhout, J.; van Vlierberghe, S.; Dubruel, P. A review of trends and limitations in hydrogel-rapid prototyping for tissue engineering. *Biomaterials* **2012**, *33*, 6020–6041. [CrossRef] [PubMed]
9. Hoffman, A.S. Hydrogels for biomedical applications. *Adv. Drug Deliv. Rev.* **2012**, *64*, 18–23. [CrossRef]
10. Wang, T.; Jiao, Y.; Chai, Q.; Yu, X. Gold Nanoparticles: Synthesis and Biological Applications. *Nano LIFE* **2015**, *5*, 1542007. [CrossRef]
11. Ding, R.; Yu, X.; Wang, P.; Zhang, J.; Zhou, Y.; Cao, X.; Tang, H.; Ayres, N.; Zhang, P. Hybrid photosensitizer based on amphiphilic block copolymer stabilized silver nanoparticles for highly efficient photodynamic inactivation of bacteria. *RSC Adv.* **2016**, *6*, 20392–20398. [CrossRef]
12. Lee, K.Y.; Mooney, D.J. Hydrogels for tissue engineering. *Chem. Rev.* **2001**, *101*, 1869–1880. [CrossRef] [PubMed]
13. Pan, L.; Yu, G.; Zhai, D.; Lee, H.R.; Zhao, W.; Liu, N.; Wang, H.; Tee, B.C.K.; Shi, Y.; Cui, Y.; et al. Hierarchical nanostructured conducting polymer hydrogel with high electrochemical activity. *Proc. Natl. Acad. Sci. USA* **2012**, *109*, 9287–9292. [CrossRef] [PubMed]

14. Zhai, D.; Liu, B.; Shi, Y.; Pan, L.; Wang, Y.; Li, W.; Zhang, R.; Yu, G. Highly sensitive glucose sensor based on Pt nanoparticle/polyaniline hydrogel heterostructures. *ACS Nano* **2013**, *7*, 3540–3546. [CrossRef] [PubMed]

15. Li, L.; Wang, Y.; Pan, L.; Shi, Y.; Cheng, W.; Shi, Y.; Yu, G. A nanostructured conductive hydrogels-based biosensor platform for human metabolite detection. *Nano Lett.* **2015**, *15*, 1146–1151. [CrossRef] [PubMed]

16. Zhu, Z.; Guan, Z.; Jia, S.; Lei, Z.; Lin, S.; Zhang, H.; Ma, Y.; Tian, Z.Q.; Yang, C.J. Au@Pt nanoparticle encapsulated target-responsive hydrogel with volumetric bar-chart chip readout for quantitative point-of-care testing. *Angew. Chem. Int. Ed.* **2014**, *53*, 12503–12507.

17. Pan, G.; Guo, Q.; Ma, Y.; Yang, H.; Li, B. Thermo-responsive hydrogel layers imprinted with RGDS peptide: A system for harvesting cell sheets. *Angew. Chem. Int. Ed.* **2013**, *52*, 6907–6911. [CrossRef] [PubMed]

18. Yu, X.; Cao, X.; Chen, X.; Ayres, N.; Zhang, P. Triplet–triplet annihilation upconversion from rationally designed polymeric emitters with tunable inter-chromophore distances. *Chem. Commun.* **2015**, *51*, 588–591. [CrossRef] [PubMed]

19. Koetting, M.C.; Guido, J.F.; Gupta, M.; Zhang, A.; Peppas, N.A. pH-responsive and enzymatically-responsive hydrogel microparticles for the oral delivery of therapeutic proteins: Effects of protein size, crosslinking density, and hydrogel degradation on protein delivery. *J. Control. Release* **2016**, *221*, 18–25. [CrossRef] [PubMed]

20. Jin, Z.; Liu, X.; Duan, S.; Yu, X.; Huang, Y.; Hayat, T.; Li, J. The adsorption of Eu(III) on carbonaceous nanofibers: Batch experiments and modeling study. *J. Mol. Liq.* **2016**, *222*, 456–462. [CrossRef]

21. Peppas, N.A.; Merrill, E.W. Poly(vinyl alcohol) hydrogels: Reinforcement of radiation-crosslinked networks by crystallization. *J. Polym. Sci. A Polym. Chem.* **1976**, *14*, 441–457. [CrossRef]

22. Matanović, M.R.; Kristl, J.; Grabnar, P.A. Thermoresponsive polymers: Insights into decisive hydrogel characteristics, mechanisms of gelation, and promising biomedical applications. *Int. J. Pharm.* **2014**, *472*, 262–275. [CrossRef] [PubMed]

23. Peppas, N. Hydrogels of poly(vinyl alcohol) and its copolymers. *Hydrogels Med. Pharm.* **1986**, *2*, 1–48.

24. Peppas, N.A.; Mongia, N.K. Ultrapure poly(vinyl alcohol) hydrogels with mucoadhesive drug delivery characteristics. *Eur. J. Pharm. Biopharm.* **1997**, *43*, 51–58. [CrossRef]

25. Hamidi, M.; Azadi, A.; Rafiei, P. Hydrogel nanoparticles in drug delivery. *Adv. Drug Deliv. Rev.* **2008**, *60*, 1638–1649. [CrossRef] [PubMed]

26. Bajpai, A.K.; Shukla, S.K.; Bhanu, S.; Kankane, S. Responsive polymers in controlled drug delivery. *Prog. Polym. Sci.* **2008**, *33*, 1088–1118. [CrossRef]

27. Overstreet, D.J.; McLemore, R.Y.; Doan, B.D.; Farag, A.; Vernon, B.L. Temperature-responsive graft copolymer hydrogels for controlled swelling and drug delivery. *Soft Matter* **2013**, *11*, 294–304. [CrossRef]

28. Zhou, Y.; Cai, Y.; Hu, X.; Long, Y. Temperature-responsive hydrogel with ultra-large solar modulation and high luminous transmission for "smart window" applications. *J. Mater. Chem. A* **2014**, *2*, 13550–13555. [CrossRef]

29. Schoener, C.A.; Hutson, H.N.; Peppas, N.A. pH-responsive hydrogels with dispersed hydrophobic nanoparticles for the oral delivery of chemotherapeutics. *J. Biomed. Mater. Res. A* **2013**, *101*, 2229–2236. [CrossRef] [PubMed]

30. De, S.K.; Aluru, N.; Johnson, B.; Crone, W.; Beebe, D.J.; Moore, J. Equilibrium swelling and kinetics of pH-responsive hydrogels: Models, experiments, and simulations. *J. Microelectromech. Syst.* **2002**, *11*, 544–555. [CrossRef]

31. Jeong, B.; Bae, Y.H.; Lee, D.S.; Kim, S.W. Biodegradable block copolymers as injectable drug-delivery systems. *Nature* **1997**, *388*, 860–862. [PubMed]

32. Dong, L.; Jiang, H. Autonomous microfluidics with stimuli-responsive hydrogels. *Soft Matter* **2007**, *3*, 1223–1230. [CrossRef]

33. Holtz, J.H.; Asher, S.A. Polymerized colloidal crystal hydrogel films as intelligent chemical sensing materials. *Nature* **1997**, *389*, 829–832. [CrossRef]

34. Zhao, Y.-L.; Stoddart, J.F. Azobenzene-Based Light-Responsive Hydrogel System. *Langmuir* **2009**, *25*, 8442–8446. [CrossRef] [PubMed]

35. Lo, C.-W.; Zhu, D.; Jiang, H. An infrared-light responsive graphene-oxide incorporated poly(N-isopropylacrylamide) hydrogel nanocomposite. *Soft Matter* **2011**, *7*, 5604–5609. [CrossRef]

36. Murakami, Y.; Maeda, M. DNA-responsive hydrogels that can shrink or swell. *Biomacromolecules* **2005**, *6*, 2927–2929. [CrossRef] [PubMed]

37. Flory, P.J.; Rehner, J., Jr. Statistical mechanics of cross-linked polymer networks II. Swelling. *J. Chem. Phys.* **1943**, *11*, 521–526. [CrossRef]

38. Flory, P.J. *Principles of Polymer Chemistry*; Cornell University Press: Ithaca, NY, USA, 1953.

39. Peppas, N.A.; Merrill, E.W. Crosslinked poly(vinyl alcohol) hydrogels as swollen elastic networks. *J. Appl. Polym. Sci.* **1977**, *21*, 1763–1770. [CrossRef]

40. Treloar, L.R.G. *The Physics of Rubber Elasticity*; Oxford University Press: Oxford, UK, 1975.

41. Flory, P.J.; Rabjohn, N.; Shaffer, M.C. Dependence of elastic properties of vulcanized rubber on the degree of cross linking. *J. Polym. Sci.* **1949**, *4*, 225–245. [CrossRef]

42. Peppas, N.; Bures, P.; Leobandung, W.; Ichikawa, H. Hydrogels in pharmaceutical formulations. *Eur. J. Pharm. Biopharm.* **2000**, *50*, 27–46. [CrossRef]

43. Lowman, A.M.; Peppas, N.A. Analysis of the complexation/decomplexation phenomena in graft copolymer networks. *Macromolecules* **1997**, *30*, 4959–4965. [CrossRef]

44. Mark, J.E. *Polymer Networks*; Springer: Heidelberg, Germany, 1982; pp. 1–26.

45. Klouda, L.; Mikos, A.G. Thermoresponsive hydrogels in biomedical applications. *Eur. J. Pharm. Biopharm.* **2008**, *68*, 34–45. [CrossRef] [PubMed]

46. Qiu, Y.; Park, K. Environment-sensitive hydrogels for drug delivery. *Adv Drug Deliver Rev.* **2001**, *53*, 321–339. [CrossRef]

47. Gao, X.; Cao, Y.; Song, X.; Zhang, Z.; Xiao, C.; He, C.; Chen, X. pH-and thermo-responsive poly(*N*-isopropylacrylamide-co-acrylic acid derivative) copolymers and hydrogels with LCST dependent on pH and alkyl side groups. *J. Mater. Chem. B* **2013**, *1*, 5578–5587. [CrossRef]

48. Schild, H. Poly(*N*-isopropylacrylamide): Experiment, theory and application. *Prog. Polym. Sci.* **1992**, *17*, 163–249. [CrossRef]

49. Ruel-Gariépy, E.; Leroux, J.-C. In situ-forming hydrogels—Review of temperature-sensitive systems. *Eur. J. Pharm. Biopharm.* **2004**, *58*, 409–426. [CrossRef] [PubMed]

50. Mortensen, K.; Pedersen, J.S. Structural study on the micelle formation of poly(ethylene oxide)-poly (propylene oxide)-poly (ethylene oxide) triblock copolymer in aqueous solution. *Macromolecules* **1993**, *26*, 805–812. [CrossRef]

51. Jaeger, J.A. Landscape division, splitting index, and effective mesh size: New measures of landscape fragmentation. *Landsc. Ecol.* **2000**, *15*, 115–130. [CrossRef]

52. Canal, T.; Peppas, N.A. Correlation between mesh size and equilibrium degree of swelling of polymeric networks. *J. Biomed. Mater. Res.* **1989**, *23*, 1183–1193. [CrossRef] [PubMed]

53. Cheng, Y.; Luo, X.; Payne, G.F.; Rubloff, G.W. Biofabrication: Programmable assembly of polysaccharide hydrogels in microfluidics as biocompatible scaffolds. *J. Mater. Chem.* **2012**, *22*, 7659–7666. [CrossRef]

54. Klemm, D.; Kramer, F.; Moritz, S.; Lindström, T.; Ankerfors, M.; Gray, D.; Dorris, A. Nanocelluloses: A New Family of Nature-Based Materials. *Angew. Chem. Int. Ed.* **2011**, *50*, 5438–5466. [CrossRef] [PubMed]

55. Bhattacharya, M.; Malinen, M.M.; Lauren, P.; Lou, Y.-R.; Kuisma, S.W.; Kanninen, L.; Lille, M.; Corlu, A.; GuGuen-Guillouzo, C.; Ikkala, O. Nanofibrillar cellulose hydrogel promotes three-dimensional liver cell culture. *J. Control. Release* **2012**, *164*, 291–298. [CrossRef] [PubMed]

56. Isobe, N.; Lee, D.-S.; Kwon, Y.-J.; Kimura, S.; Kuga, S.; Wada, M.; Kim, U.-J. Immobilization of protein on cellulose hydrogel. *Cellulose* **2011**, *18*, 1251–1256. [CrossRef]

57. Kim, M.H.; An, S.; Won, K.; Kim, H.J.; Lee, S.H. Entrapment of enzymes into cellulose–biopolymer composite hydrogel beads using biocompatible ionic liquid. *J. Mol. Catal. B Enzym.* **2012**, *75*, 68–72. [CrossRef]

58. Bott, K.; Upton, Z.; Schrobback, K.; Ehrbar, M.; Hubbell, J.A.; Lutolf, M.P.; Rizzi, S.C. The effect of matrix characteristics on fibroblast proliferation in 3D gels. *Biomaterials* **2010**, *31*, 8454–8464. [CrossRef] [PubMed]

59. Nichol, J.W.; Koshy, S.T.; Bae, H.; Hwang, C.M.; Yamanlar, S.; Khademhosseini, A. Cell-laden microengineered gelatin methacrylate hydrogels. *Biomaterials* **2010**, *31*, 5536–5544. [CrossRef] [PubMed]

60. Fu, Y.; Xu, K.; Zheng, X.; Giacomin, A.J.; Mix, A.W.; Kao, W.J. 3D cell entrapment in crosslinked thiolated gelatin-poly(ethylene glycol) diacrylate hydrogels. *Biomaterials* **2012**, *33*, 48–58. [CrossRef] [PubMed]

61. Lin, R.-Z.; Chen, Y.-C.; Moreno-Luna, R.; Khademhosseini, A.; Melero-Martin, J.M. Transdermal regulation of vascular network bioengineering using a photopolymerizable methacrylated gelatin hydrogel. *Biomaterials* **2013**, *34*, 6785–6796. [CrossRef] [PubMed]

62. Pok, S.; Myers, J.D.; Madihally, S.V.; Jacot, J.G. A multilayered scaffold of a chitosan and gelatin hydrogel supported by a PCL core for cardiac tissue engineering. *Acta Biomater.* **2013**, *9*, 5630–5642. [CrossRef] [PubMed]

63. Milašinović, N.; Kalagasidis Krušić, M.; Knežević-Jugović, Z.; Filipović, J. Hydrogels of *N*-isopropylacrylamide copolymers with controlled release of a model protein. *Int. J. Pharm.* **2010**, *383*, 53–61. [CrossRef] [PubMed]

64. Anghelache, A.; Teodorescu, M.; Stić, M.; Knežević-Jugović, Z.; Filipović, J. Novel crosslinked thermoresponsive hydrogels with controlled poly(ethylene glycol)—poly(propylene glycol) multiblock copolymer structure. *Colloid Polym. Sci.* **2014**, *292*, 829–838. [CrossRef]

65. Gong, J.P.; Katsuyama, Y.; Kurokawa, T.; Osada, Y. Double-Network Hydrogels with Extremely High Mechanical Strength. *Adv. Mater.* **2003**, *15*, 1155–1158. [CrossRef]

66. Naficy, S.; Razal, J.M.; Whitten, P.G.; Wallace, G.G.; Spinks, G.M. A pH-sensitive, strong double-network hydrogel: Poly(ethylene glycol) methyl ether methacrylates–poly(acrylic acid). *J. Polym. Sci. B Polym. Phys.* **2012**, *50*, 423–430. [CrossRef]

67. Zhang, H.; Qadeer, A.; Chen, W. In situ gelable interpenetrating double network hydrogel formulated from binary components: Thiolated chitosan and oxidized dextran. *Biomacromolecules* **2011**, *12*, 1428–1437. [CrossRef] [PubMed]

68. Huang, P.; Chen, W.; Yan, L. An inorganic–organic double network hydrogel of graphene and polymer. *Nanoscale* **2013**, *5*, 6034–6039. [CrossRef] [PubMed]

69. Baroli, B. Hydrogels for tissue engineering and delivery of tissue-inducing substances. *J. Pharm. Sci.* **2007**, *96*, 2197–2223. [CrossRef] [PubMed]

70. Nicodemus, G.D.; Bryant, S.J. Cell encapsulation in biodegradable hydrogels for tissue engineering applications. *Tissue Eng. B Rev.* **2008**, *14*, 149–165. [CrossRef] [PubMed]

71. Alge, D.L.; Azagarsamy, M.A.; Donohue, D.F.; Anseth, K.S. Synthetically tractable click hydrogels for three-dimensional cell culture formed using tetrazine–norbornene chemistry. *Biomacromolecules* **2013**, *14*, 949–953. [CrossRef] [PubMed]

72. Kaemmerer, E.; Melchels, F.P.; Holzapfel, B.M.; Meckel, T.; Hutmacher, D.W.; Loessner, D. Gelatine methacrylamide-based hydrogels: An alternative three-dimensional cancer cell culture system. *Acta Biomater.* **2014**, *10*, 2551–2562. [CrossRef] [PubMed]

73. Wu, D.Y.; Meure, S.; Solomon, D. Self-healing polymeric materials: A review of recent developments. *Prog. Polym. Sci.* **2008**, *33*, 479–522. [CrossRef]

74. Thakur, V.K.; Kessler, M.R. Self-healing polymer nanocomposite materials: A review. *Polymer* **2015**, *69*, 369–383. [CrossRef]

75. Deng, G.; Li, F.; Yu, H.; Liu, F.; Liu, C.; Sun, W.; Jiang, H.; Chen, Y. Dynamic hydrogels with an environmental adaptive self-healing ability and dual responsive sol–gel transitions. *ACS Macro Lett.* **2012**, *1*, 275–279. [CrossRef]

76. Amiri, S.; Rahimi, A. Hybrid nanocomposite coating by sol–gel method: A review. *Iran Polym. J.* **2016**, *25*, 559–577. [CrossRef]

77. Bode, S.; Bose, R.; Matthes, S.; Ehrhardt, M.; Seifert, A.; Schacher, F.; Paulus, R.; Stumpf, S.; Sandmann, B.; Vitz, J. Self-healing metallopolymers based on cadmium bis(terpyridine) complex containing polymer networks. *Polym. Chem.* **2013**, *4*, 4966–4973. [CrossRef]

78. Burattini, S.; Colquhoun, H.M.; Fox, J.D.; Friedmann, D.; Greenland, B.W.; Harris, P.J.; Hayes, W.; Mackay, M.E.; Rowan, S.J. A Self-repairing, supramolecular polymer system: Healability as a consequence of donor–acceptor π–π stacking interactions. *Chem. Commun.* **2009**, 6717–6719. [CrossRef] [PubMed]

79. Mukherjee, S.; Hill, M.R.; Sumerlin, B.S. Self-healing hydrogels containing reversible oxime crosslinks. *Soft Matter* **2015**, *11*, 6152–6161. [CrossRef] [PubMed]

80. Tuncaboylu, D.C.; Sari, M.; Oppermann, W.; Okay, O. Tough and self-healing hydrogels formed via hydrophobic interactions. *Macromolecules* **2011**, *44*, 4997–5005. [CrossRef]

81. Tamesue, S.; Yasuda, K.; Noguchi, S.; Mitsumata, T.; Yamauchi, T. Highly Tolerant and Durable Adhesion between Hydrogels Utilizing Intercalation of Cationic Substituents into Layered Inorganic Compounds. *ACS Macro Lett.* **2016**, *5*, 704–708. [CrossRef]

82. Gupta, P.; Vermani, K.; Garg, S. Hydrogels: From controlled release to pH-responsive drug delivery. *Drug Discov. Today* **2002**, *7*, 569–579. [CrossRef]

83. Hoare, T.R.; Kohane, D.S. Hydrogels in drug delivery: Progress and challenges. *Polymer* **2008**, *49*, 1993–2007. [CrossRef]

84. Paavola, A.; Yliruusi, J.; Kajimoto, Y.; Kalso, E.; Wahlström, T.; Rosenberg, P. Controlled release of lidocaine from injectable gels and efficacy in rat sciatic nerve block. *Pharm. Res.* **1995**, *12*, 1997–2002. [CrossRef] [PubMed]

85. Sosnik, A.; Cohn, D. Ethoxysilane-capped PEO–PPO–PEO triblocks: A new family of reverse thermo-responsive polymers. *Biomaterials* **2004**, *25*, 2851–2858. [CrossRef] [PubMed]

86. Cho, K.Y.; Chung, T.W.; Kim, B.C.; Kim, M.K.; Lee, J.H.; Wee, W.R.; Cho, C.S. Release of ciprofloxacin from poloxamer-graft-hyaluronic acid hydrogels in vitro. *Int. J. Pharm.* **2003**, *260*, 83–91. [CrossRef]

87. Kim, M.R.; Park, T.G. Temperature-responsive and degradable hyaluronic acid/Pluronic composite hydrogels for controlled release of human growth hormone. *J. Control. Release* **2002**, *80*, 69–77. [CrossRef]

88. Xiong, L.; Luo, Q.; Wang, Y.; Li, X.; Shen, Z.; Zhu, W. An injectable drug-loaded hydrogel based on a supramolecular polymeric prodrug. *Chem. Commun.* **2015**, *51*, 14644–14647. [CrossRef] [PubMed]

*gels*

*Article*

# A Bioactive Hydrogel and 3D Printed Polycaprolactone System for Bone Tissue Engineering

Ivan Hernandez [1], Alok Kumar [1,*] and Binata Joddar [1,2]

1   Inspired Materials & Stem-Cell Based Tissue Engineering Laboratory (IMSTEL),
    Department of Metallurgical, Materials and Biomedical Engineering, University of Texas at El Paso,
    El Paso, TX 79968, USA; ihernandez38@miners.utep.edu
2   Border Biomedical Research Center, University of Texas at El Paso, El Paso, TX 79968, USA;
    bjoddar@utep.edu
*   Correspondence: akumar3@utep.edu

Received: 27 May 2017; Accepted: 4 July 2017; Published: 6 July 2017

**Abstract:** In this study, a hybrid system consisting of 3D printed polycaprolactone (PCL) filled with hydrogel was developed as an application for reconstruction of long bone defects, which are innately difficult to repair due to large missing segments of bone. A 3D printed gyroid scaffold of PCL allowed a larger amount of hydrogel to be loaded within the scaffolds as compared to 3D printed mesh and honeycomb scaffolds of similar volumes and strut thicknesses. The hydrogel was a mixture of alginate, gelatin, and nano-hydroxyapatite, infiltrated with human mesenchymal stem cells (hMSC) to enhance the osteoconductivity and biocompatibility of the system. Adhesion and viability of hMSC in the PCL/hydrogel system confirmed its cytocompatibility. Biomineralization tests in simulated body fluid (SBF) showed the nucleation and growth of apatite crystals, which confirmed the bioactivity of the PCL/hydrogel system. Moreover, dissolution studies, in SBF revealed a sustained dissolution of the hydrogel with time. Overall, the present study provides a new approach in bone tissue engineering to repair bone defects with a bioactive hybrid system consisting of a polymeric scaffold, hydrogel, and hMSC.

**Keywords:** 3D printing; polycaprolactone (PCL); hydroxyapatite; hydrogel; bone defect

---

## 1. Introduction

Traditionally, bone fractures and defects created due to injury or disease are treated by temporary and/or permanent implants [1]. However, inadequate bone growth leads to non-union of newly formed bone in such circumstances [2,3]. Therefore, a major concern in repairing bone defects is absence of a suitable implant to accelerate bone regeneration and induce bone union. Alternatively, in bone-tissue engineering, biomaterials alone or in combination with suitable biological and chemical factors are used to restore the functionality of injured bone tissue [4]. In this context, implantation of cell-seeded scaffold constructs have been used to enhance bone regeneration [5,6]. Furthermore, presence of a bioactive biomaterial, such as autologous bone harvested from the patient's own body or osteogenic supplements can be helpful in bone formation [7,8]. Therefore, in the past, several methods have been developed for delivering osteoblasts (bone forming cells) and osteogenic growth factors at defect sites [8–10]. Concurrently, three dimensional (3D) printing methods have been used to create uniquely designed scaffolds for faster recovery from bone injuries [11]. 3D printed porous scaffolds with interconnected pores should allow the formation of vascularized tissue, which is required to supply nutrients and oxygen to growing cells inside the pores [12]. Unlike very expensive additive manufacturing methods such as electron beam melting (EBM) and selective laser sintering (SLS) used to fabricate high strength scaffolds, 3D printing methods such as fused deposition modeling (FDM) can alternatively be used to fabricate scaffolds at much lower costs [11].

Among various characteristics required for reconstruction of bone defects, a patient-specific design mimicking the fractured bone, with an ability to promote bone ingrowth and healing is required for faster recovery from bone-injuries. Therefore, in this study, we developed a hybrid system of a 3D printed scaffold of polycaprolactone (PCL) and bioactive hydrogel infiltrated with human mesenchymal stem cells (hMSC). Our overall objective was to develop an on-demand method to provide a support system for application in segmental bone defect restoration with a custom-made PCL scaffold, which would also deliver a bioactive hydrogel and hMSC for induction of bone growth. Furthermore, this PCL/hydrogel system would not be cytotoxic and would support bone repair as it contained hMSC. The novelty in this study is the simultaneous application of both a hydrogel (with hMSC) and a 3D printed PCL scaffold to make a hybrid system with bioactive properties and capability to support and maintain the structural integrity of bone, during repair and regeneration.

## 2. Results

### 2.1. Hydrogel Preparation

The hydrogel was prepared by the crosslinking of an alginate and gelatin mixture with 1-ethyl-3-(3-dimethylaminopropyl) carbodiimide (EDC) and *N*-hydroxy-succinimide (NHS), followed by calcium chloride (CaCl$_2$). During hydrogel synthesis, EDC was used to activate the carboxyl groups of alginate to form active ester groups, followed by NHS bonding with alginate due to replacement of EDC to improve the efficiency of amine reaction [13,14]. To synthesize the pre-hydrogel (Figure S1), NHS activated carboxylic group of alginate was linked with the primary amine of gelatin by replacing NHS. Addition of CaCl$_2$ led to the ionic interactions of α-L-guluronic acid (G-block) of alginate in the pre-hydrogel to form a stable crosslinked hydrogel. Cross-linking with CaCl$_2$ was done to enhance the retention of the hydrogel within the PCL scaffold structure (Figure S2).

### 2.2. Rapid Fabrication of the 3D Printed PCL Scaffold

Using FDM technology, primarily, a cylindrical-shaped gyroid PCL scaffold (height: 5 mm, diameter: 15 mm) was printed within 10 min (Figure S3). Mesh and honeycomb structures of similar dimensions were also printed to compare the hydrogel retention capacity of scaffolds.

### 2.3. Hydrogel Retention Capacity of the PCL Scaffolds

As shown in Figure 1, three different 3D printed structures: mesh, honeycomb, and gyroid of identical dimensions were filled with hydrogel to compare the irrespective gel loading capabilities. Results showed an average of 1.25 ± 0.04 g, 0.82 ± 0.04 g and 0.46 ± 0.04 g of hydrogel loaded in the gyroid, mesh, and honeycomb structures, respectively. The amount of hydrogel loaded within the gyroid was about ~35% greater compared to mesh and ~63% greater compared to honeycomb, respectively. Since the 3D PCL scaffold took less than 10 min to print, this composite PCL/hydrogel implant can be applied to an in vivo application within a desired and short time frame if needed. Furthermore, we noted a higher amount of hydrogel retention in gyroid structure than in mesh and honeycomb structures. Therefore, only gyroid structure was used for further study.

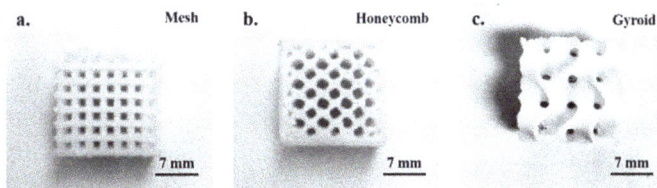

**Figure 1.** Digital pictures of 3D printed mesh (**a**), honeycomb (**b**), and gyroid (**c**) structures of identical dimensions.

## 2.4. Microstructure Imaging and Characterization of Phases in the PCL Scaffold and Hydrogel System

Scanning electron microscope (SEM) images revealed a highly porous structure of the hydrogel with an average pore size of 399.22 ± 22.03 µm (Figure 2). The PCL scaffold (gyroid) was characterized by an average strut diameter of 320.17 ± 3.47 µm. X-ray diffraction (XRD) data showed the presence of signatory diffraction peaks of hydroxyapatite (HA), alginate and gelatin in the hydrogel (Figure 3). The sharp narrow peaks of HA and PCL confirmed the crystallinity of the nano-HA and PCL.

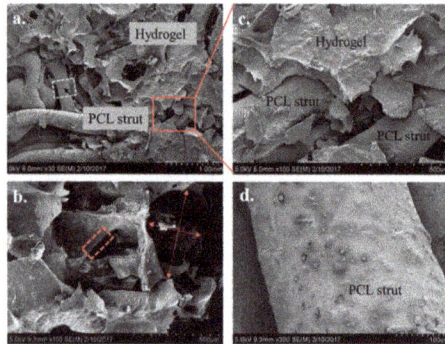

**Figure 2.** Scanning electron microscope (SEM) images of freeze-dried polycaprolactone (PCL)-gel samples (**a**). A high magnification image confirmed the highly porous nature of the hydrogel with interconnected pores. The pore shape and pore wall thickness are marked with a cross-arrow and a rectangular box, respectively (**b**). A magnified image of region marked with rectangular box in (**a**) showed complete adherence of hydrogel on the scaffold, which is expected to provide a bioactive coating to the otherwise bioinert surface of PCL (**c**). The PCL scaffold was characterized by surface micro-roughness and non-homogeneity (**d**).

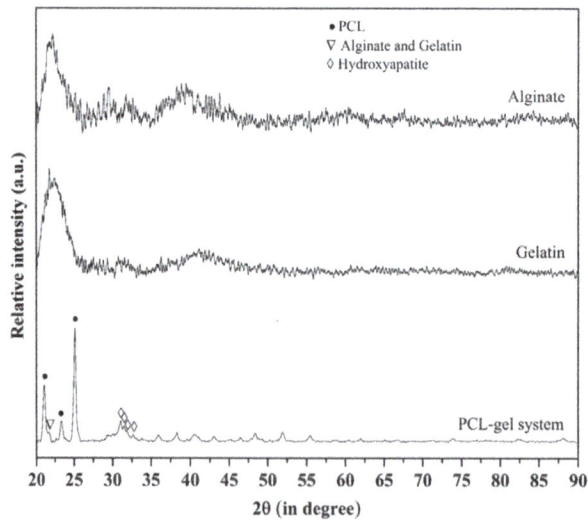

**Figure 3.** A comparison of X-ray diffraction (XRD) data of hybrid PCL/hydrogel scaffolds with alginate, and gelatin confirmed the presence of semi-crystalline phases of alginate and gelatin in the hydrogel loaded in the PCL scaffold (∇). The diffraction data also confirmed the presence of PCL (●) and hydroxyapatite (HA) (◊) in its monolithic phase.

## 2.5. Sustained Dissolution of Hydrogel in Simulated Body Fluid (SBF)

Constant visual monitoring of the samples during and after the completion of the dissolution study in SBF showed a uniform dissolution of hydrogel during the 12 days of test period. After the test, optical density of the spent SBF was measured to estimate the amount of dissolved hydrogel (Figure 4). This measured optical density was converted to actual amounts of hydrogel dissolved per unit volume of spent SBF, based on a standard curve (Figure S4). Results revealed a continuous dissolution profile for hydrogel from day 1–6 of the study. An average of 12.37 ± 0.90 mg hydrogel dissolved in the first three days, followed by 5.58 ± 0.16 mg hydrogel dissolution in next 3 days. After 6 days, a decrease in the rate of dissolution was noted with only 2.1 ± 0.32 mg hydrogel found dissolved in the next 6 days.

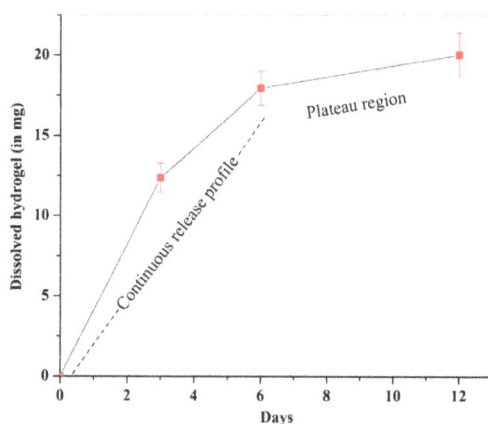

**Figure 4.** The dissolution study carried out in simulated body fluid (SBF) for 3, 6, and 12 days showed the continuous dissolution of hydrogel with time, with decrease in dissolution rate after 3 days. A plateau region after 6 days can either be associated with significant decrease in degradation rate of hydrogel or predominant apatite deposition from the SBF (see Figure 5).

## 2.6. Apatite Formation Ability of the PCL/Gel System

The in vitro apatite formation ability of a biomaterial can be correlated to its in vivo bone-bonding ability [15]. In this context, samples used in the dissolution study in SBF were further used to study the formation and dissolution of apatite within these PCL-gel samples. SEM images showed a uniform deposition of apatite with higher amounts deposited within the hydrogel as compared to PCL scaffolds (Figure 5). After 3 days of immersion in SBF, deposition of apatite crystals was noted. After 6 and 12 days (Figure 5c,d, respectively), a thick layer of apatite was noted due to the continuous deposition of apatite. A strain-induced crack, resulting from drying of the samples was used to estimate the apatite layer thickness after 12 days, which revealed a thickness of 15.62 ± 0.51 µm (Figure S5). Higher magnification SEM micrographs showed a higher amount of deposited apatite after 3 days on PCL struts as compared to after 6 and 12 days (Figure 6a–c). In contrast, as shown in Figure 6d–f, more apatite formation with time was observed with a relatively denser apatite layer noted on day 12. Higher magnification images (Figure 6g–i) showed a change in morphology of the deposited apatite with time. After 3 days, a globular morphology of apatite was noted on the hydrogel, which changed to acicular morphology on day 6. After 12 days, a denser apatite layer was observed with rod-like fine spherical-shaped apatite crystals.

In summary, time dependent apatite formation and stabilization was noted within the hydrogel with a higher amount of apatite on day 12 as compared to days 3 and 6. However, a smaller amount of apatite was formed on the PCL surface which was reduced with time due to noticeable amounts of dissolution of apatite layer from the PCL surfaces.

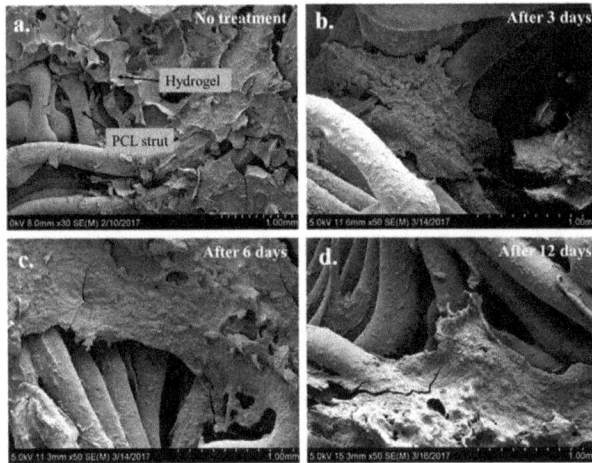

**Figure 5.** Low magnification SEM images of freeze-dried PCL-gel samples without SBF (**a**) and with SBF treatment for 3 (**b**), 6 (**c**), and 12 days (**d**). The SBF treated samples showed homogenous apatite layer over the hydrogel as well as PCL struts with an increasing amount of apatite deposition with time. A crack in apatite layer in (**c,d**) is due the strain generated due to drying of the samples.

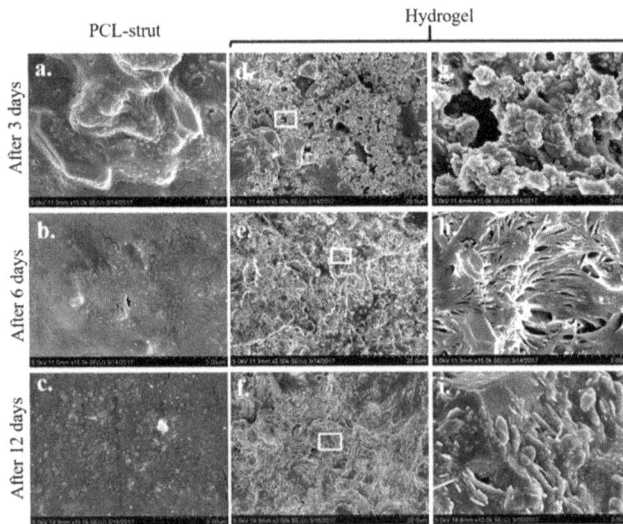

**Figure 6.** High magnification SEM images of freeze-dried PCL/ hydrogel samples after 3 (**a,d,g**), 6 (**b,e,h**), and 12 days (**c,f,i**) of immersion in SBF. The (**g**), (**h**), and (**i**) are the magnified images of regions marked in micrographs (**d**), (**e**), and (**f**), respectively. Results showed the deposition of apatite on both PCL as well hydrogel (**a,d**) in the initial period (3 days) of SBF immersion. A lower amount of apatite on PCL struts than hydrogel after 6 and 12 days may be due to the dissolution of deposited apatite from PCL. Scale bar for (**a–c,g–i**) is 3 μm and for (**d–f**) 20 μm.

## 2.7. Cytocompatibility of the PCL/Gel System

Figure 7a showed the presence of viable pre-stained hMSC in the entire system, although there were more cells in the hydrogel than on the scaffold strut (Figure 7b). Since PCL is bioinert in nature and does not support cell adhesion [16], a bioactive hydrogel infiltrated with hMSC was loaded within the pores of the scaffold. Due to the bioactive property of the hydrogel, we expected a higher number of viable cells to be retained within the hydrogel. Figure 7c showed a large number of cells with elliptical-shaped morphology (Figure 7d). Control samples did not fluoresce at all (Figure S6), confirming the presence of viable hMSC in the samples imaged and reported in this study.

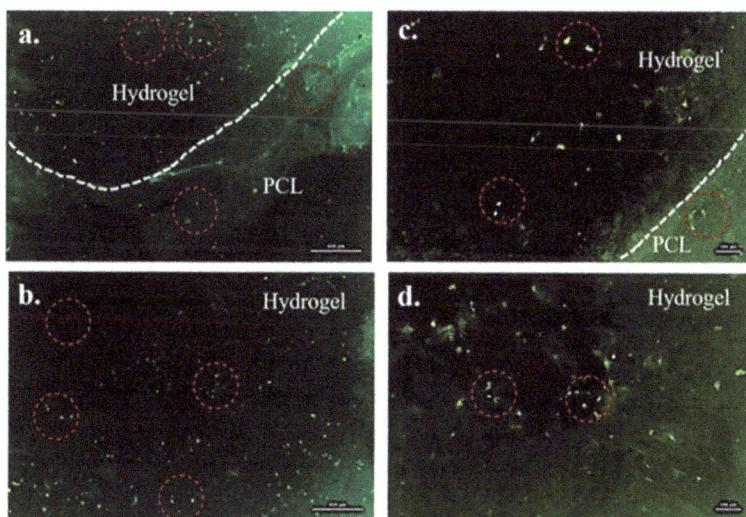

**Figure 7.** Representative fluorescence images of PCL-gel samples seeded with pre-stained human mesenchymal stem cells showed the presence of cells (green) in the hydrogel (**a,b**) as well as on the PCL struts (**a**). The white-colored broken line shows the boundary between the PCL scaffold and hydrogel. The cells are marked with red circles within both the hydrogel and scaffold areas. Images (**c,d**) are the magnified images of micrographs (**a,b**), respectively. Scale bar for (**a,b**) is 500 µm and for (**c,d**) is 100 µm.

## 3. Discussion

Hydrogels have been widely used for most tissue engineering applications [17]. Specifically, hydrogel allows higher cell encapsulation and therefore, delivery of a greater concentration of cells at the site of implantation/defect, which in turn could accelerate the regeneration of the damaged tissue [18,19]. However, they possess weak mechanical properties that can be optimized by functionalization or crosslinking [20,21], or by serving as a bioactive filler material within a bioinert scaffold, which by itself does not interact with the body [22]. However, such a scaffold can still be structurally capable of supporting cell-growth, cell-proliferation, and vascularization [23]. The idea of using a supporting scaffold and a bioactive filler material with cells was based on the work by Gugala et al. [7], in which a bioactive graft material and a porous support structure was used to restore a bone defect in vivo. No healing was noted in the absence of any support material. But, a partial healing was found in case of perforated polylactic acid sheet used as a scaffold [7]. In contrast to this, complete healing was achieved when this scaffold was filled with autogenous cancellous bone [7]. In summary, this study emphasized the importance of both a support scaffold and a bioactive material to promote bone regeneration [7]. Considering the importance of the role played by the presence of osteogenic material in new bone formation, in the present study a novel approach was explored to incorporate hMSC and HA in an alginate and gelatin-based hydrogel and filled within a 3D printed PCL scaffold. Multipotent stromal hMSC were added to improve the osteoconductive properties of the hydrogel [24]. Bone and teeth of most animals, including humans, are composed of calcium phosphate (e.g., HA) which makes up 62–65% of the total bone composition [25]. Calcium phosphates have intrinsic properties that stimulate bone regeneration [26–28]. Therefore, presence of calcium phosphates, such as HA, is expected to improve the bioactivity of the designed PCL-gel samples [28–30]. Alginate is a hydrophilic anionic polysaccharide and exhibits chelation in the presence of divalent cations such as $Ca^{2+}$ and $Mg^{2+}$ [31]. Since alginate is unable to interact with cells, gelatin was added to improve cell adhesion [32,33].

Since hydrogels are mechanically fragile, to improve its retention at the application site, the hydrogel was loaded in a 3D printed gyroid PCL and post-crosslinked. Furthermore, a gyroid structure is characterized as having a minimal surface area and architectural as well as mechanical characteristics similar to trabecular bone [34–36]. The highest amount of hydrogel loading capacity in the gyroid scaffolds could be related to the larger pore size and minimal surface area as compared to other scaffold designs [35]. This network of larger sized channels can facilitate invasion of the host vasculature post in vivo implantation [37].

Apatite nucleation and growth is a dynamic process and depends on the concentration of calcium and phosphate ions in the SBF [28]. The entire process of apatite growth on biodegradable calcium phosphate-based biomaterials can be divided into two steps [38]. During the first step, the biomaterial (hydrogel) dissolved and supplemented the SBF with calcium and phosphate ions until super-saturation was achieved [39]. In the second step, calcium and phosphate ions started depositing on the biomaterial surface from the supersaturated solution [39]. In the case of PCL, decline in the amounts of apatite could be correlated to the bioinertness of PCL [40]. In contrast to this, an increase in apatite on the hydrogel with time was due to the bioactive nature of the hydrogel [41]. It is known that a dissolution study in SBF can simulate in vivo physiological conditions allowing dynamic interplay of material dissolution accompanied by bone mineralization and deposition [42]. Since the designed PCL-gel system is bioactive, it can support the osteogenic differentiation of hMSC environment [43,44] and thus, can accelerate osseointegration [45] when applied in vivo.

Although the primary objective of this study was to develop a bioactive and biodegradable hybrid system of a polymeric scaffold and hydrogel, we also aimed to create an easy and rapid system for on-demand applications in bone tissue engineering. A custom-made 3D printed reconstruct implant specifically designed to fit in the defect can minimize the micromotion of the implant at the host site and enable firmly-anchored new bone formation at the interface of implant and bone. In addition, a rapid method for fabrication of a 3D printed scaffold–hydrogel hybrid system can be used to reconstruct and stabilize architecturally complex bone fractures [11].

In summary, the designed hybrid system of a PCL scaffold and bioactive hydrogel was cytocompatible with an ability to promote apatite formation. Therefore, the designed PCL-gel system can potentially be used to repair custom-sized bone defects. However, further studies are required to test the efficacy of this system for promotion of vascularization and osseointegration.

## 4. Conclusions

The hybrid PCL-gel system indicated good cytocompatibility, showing adhesion and viability of the hMSC within the hydrogel matrix as well as on the solid scaffold surfaces. Further, the biomineralization test in SBF showed the nucleation and growth of apatite crystals on the hydrogel as well as the PCL scaffold, which confirmed its bioactivity. This hybrid PCL-gel system can be optimized to fabricate an implantable device within a short time to provide on-demand patient-specific solutions. Overall, the present study provides a new approach in bone tissue engineering for repair of bone defects, with a bioactive hybrid system of a biodegradable scaffold and hydrogel. Furthermore, in vitro studies could be carried out in future to combine endothelial cells and growth factors in addition to hMSC to induce vascularization. Also, in vivo studies could be carried out to study the effect of dissolution rate of the implant and new bone formation on the overall bone repair process.

## 5. Materials and Methods

### 5.1. Materials

Sodium alginate (Cat. No. 218295) and type-A gelatin (Cat. No. 901771) was obtained from MP Biomedicals (Strasbourg, France). 1-ethyl-3-(-3-dimethylaminopropyl) carbodiimide hydrochloride (EDC, Cat. No. 22980) and *N*-hydroxysuccinimide (NHS, Cat. No. 24500) were purchased from Thermo Scientific (Rockford, IL, USA). Calcium chloride (Cat. No. C79-500) was purchased from Fisher Chemicals (Fair Lawn, NJ, USA). Nanocrystalline HA was synthesized using a previously reported method of suspension-precipitation with calcium oxide and orthophosphoric acid added as a precursor [16]. HA powder was prepared during one of our previous study [47] and was used in this study without any modification. HA particles were characterized by length and width of $120 \pm 38$ nm and $52 \pm 25$ nm, respectively [47]. PCL (Cat. No. B01M8IDB07) filament with 1.75 mm diameter was obtained from Shenzhen Esun Industrial Co. (Shenzhen, China). In addition, hMSC (Cat. No. SV30110.01), basal culture medium (Cat. No. SH30879.01), and growth supplement (Cat. No. SH30878.01) were obtained from HyClone Laboratories, GE Healthcare Life Sciences (Logan, UT, USA). A green florescent dye (PKH67, Cat. No. MINI67) for the pre-staining of cells prior to cell culture was purchased from the Sigma Aldrich (St. Louis, MO, USA). $1\times$ Cell Dissociation Medium (0.25%Trypsin supplemented with 2.21 mM EDTA, Cat. No. 25-053-Cl) and $1\times$ Phosphate Buffered Saline ($1\times$ PBS, Cat. No. K812-500) were purchased from Mediatech, Corning (Masassas, VA, USA) and Amresco (Solon, OH, USA), respectively. A sterilized syringe of 10 mL volume (Cat. No. DG515805) and a needle of 0.7 mm inner diameter (Cat. No. B01KZ0MHSC) were procured from Becton Dickinson (Franklin lakes, NJ, USA) and Huaha (China), respectively. For the bioactivity study, $1\times$ simulated body fluid ($1\times$ SBF) was prepared according to the method described by Oyane et al. [48] and has been reported elsewhere [28].

### 5.2. 3D Printing of PCL Scaffold

In this study, cylindrical-shaped (height: 5 mm, diameter: 15 mm) PCL scaffolds of gyroid structure (65% porosity and 1.2 mm pore size) were printed using a Fused Deposition Modeling (FDM) printer (MakerBot, New York City, NY, USA, Model. Replicator Mini 5th generation) (Figure S3). Printing was accomplished with a nozzle diameter of 0.4 mm, operating temperature of 110 °C, and a printing speed of 90 mm/s with 100% material filling density. Prior to use, the scaffolds were washed in distilled water and sterilized by soaking in 70% ethanol for 20 min, followed by UV irradiation for 30 min. The extent of porosity of these scaffolds was measured by comparing the weight of porous scaffolds with solid scaffolds of similar dimensions.

### 5.3. Synthesis of Bioactive Hydrogel Infiltrated with hMSC

The method used for hydrogel synthesis was adopted from previous studies carried out by Wang et al. [49]. Briefly, 50 mg nanocrystalline powder of HA was added in 10 mL distilled water and mixed by magnetic stirring (at 100 rpm for 15 min at room temperature). Next, gelatin and sodium alginate, 200 mg of each were added and stirred again (at room temperature for 15 min at 100 rpm). After this, 25 mg EDC was added (stirred at room temperature for 10 min), followed by addition of 15 mg NHS (stirring for another 5 min at room temperature) to make the hydrogel mixture. For sterilization, this hydrogel was irradiated with UV light for 30 min. After sterilization, hMSC were mixed with the hydrogel. For doing this, first, hMSC were cultured in a T-75 flask, until ~70% confluency was reached. ~70% confluent layer of hMSC was trypsinized using trypsin-EDTA and cells (cell density $\sim 2 \times 10^7$ cells/mL) were pre-stained with PKH67 as per manufacturer's protocols. These pre-stained cells were centrifuged to remove the cell-suspension media and mixed with hydrogel. Hydrogel mixed with hMSC was loaded in a sterilized syringe and was injected in the pores of scaffolds. A detailed method of hydrogel loading in the scaffold is provided in the following Section 5.4.

## 5.4. Formation of a Hybrid PCL/Hydrogel System

The cell-loaded hydrogel was filled in a sterilized syringe and injected in the 3D printed porous PCL gyroid scaffold. This "hybrid PCL/hydrogel system" was addressed as "PCL-gel sample" thereafter, was treated with 1 M $CaCl_2$ for 10 min to prevent the leakage of hydrogel from the scaffold, followed by washing with 1× PBS for 10 min. Furthermore, PCL-gel samples were washed twice for 5 min each using complete culture medium (basal medium with 10% growth supplement). These PCL-gel samples with hMSC were then transferred to 24 well-plate and incubated in the presence of complete culture medium (5% $CO_2$ and 95% relative humidity at 37 °C). The details of cell culture and viability analysis are reported in Section 5.5.

The gyroid structure was compared with commonly used mesh and honeycomb structures for its efficacy to allow the high loading of hydrogel within the pores of scaffolds. For this, the weight of 3D printed scaffolds of identical size and shape was measured before and after the hydrogel (without cells) loading in the scaffolds. The difference in weight was used to estimate the amount of hydrogel loaded in the scaffolds using Equation (1). In addition to this, PCL-gel samples without cells were used for the dissolution study and bioactivity test as well.

$$\text{Amount of loaded hydrogel in the scaffold (mg)} = m_h - m \tag{1}$$

where, $m_h$ and $m$ was weight of hydrogel loaded scaffold and bare scaffold, respectively.

## 5.5. Cytocompatibility Assessment

PCL-gel samples (with pre-stained hMSC in hydrogel) were transferred to a 24 well-plate and incubated in the presence of 2 mL complete culture medium for 48 h (in 5% $CO_2$ and 95% relative humidity at 37 °C). After 48 h of incubation, samples were observed under the fluorescent microscope to investigate cell adhesion, growth, and viability. At least 3 samples were used the study.

## 5.6. Dissolution Study and Bioactivity Test

To analyze the dissolution behavior and bioactivity, PCL-gel samples were immersed in 2 mL simulated body fluid (1× SBF) in 24 well-plate and incubated for 3, 6, and 12 days. Samples were incubated at 37 °C, 5% $CO_2$ and 95% relative humidity, and media was not changed during the incubation period. At least 3 samples were used in each category.

### 5.6.1. Calculation of the Dissolved Amount of Hydrogel

After completion of incubation, PCL-gel samples were carefully removed from the media and stored at −80 °C for 12 h in a closed container for SEM analysis. The detailed method of analysis is provided in the consecutive Section 5.7.

After removal of PCL-gel samples, SBF solution was transferred to wells of a 96 well-plate to record the absorption (optical density) at 630 nm using optical density reader (Model: ELx800, BioTek, Winooski, VT, USA) and data was compared with the optical density (OD) of untreated 1× SBF, used as a reference during absorption measurements.

To calculate the weight of dissolved hydrogel in SBF during incubation, first a standard curve was plotted using known amount of hydrogel, dissolved in 1× SBF (0.5, 1, 1.5 mg/mL) and corresponding OD (at 630 nm) was recorded (Figure S4). The Equation (2) obtained from the standard curve was used to estimate the amount of dissolved hydrogel in the SBF.

$$\text{Dissolved amount of hydrogel} \left(\frac{\text{mg}}{\text{mL}}\right) = \left(\frac{0.0003 + OD_s}{0.0157}\right) \tag{2}$$

where, $OD_s$ was the optical density of the SBF after 3, 6, and 12 days of immersion testing.

### 5.6.2. Apatite Formation on the PCL-Gel Samples

As mentioned in the previous section, after dissolution study, PCL-gel samples were removed from the SBF and refrigerated for 12 h at −80 °C. These refrigerated samples were then lyophilized for 12 h and after this, were analyzed using scanning electron microscope (SEM, S-4800, Hitachi, Tokyo, Japan) to study the morphology of deposited apatite on samples as well as the mechanism of apatite formation. Apatite formation data was correlated with the dissolution data to better understand the relationship between dissolution and deposition.

### 5.7. XRD and SEM Analysis

For the phase analysis, PCL-gel samples without cells were lyophilized prior to X-ray diffraction (XRD, D8 Discover, Bruker's diffractometer, Karlsruhe, Germany). XRD was carried out at 40 kV voltage and 40 mA current with CuKα wavelength (1.54056 Å) and 2θ ranges from 20° to 90° at a scanning rate of 2°/min with a step size of 0.02°.

SEM was operated in secondary electron mode for the analysis of morphology of PCL-gel samples, before and after dissolution study. Scaffolds without hydrogel were also studied and compared with the PCL-gel samples. Prior to SEM, to minimize charging during observation, samples were gold-coated using sputter coater (Model: EMS150R ES, Quorum, Laughton, East Sussex, UK), equipped with a gold/palladium target (Cat. No. 91017-AP, Electron Microscopy Sciences, Hatfield, PA, USA).

**Supplementary Materials:** The following are available online at http://www.mdpi.com/2310-2861/3/3/26/s1, Figure S1: Digital pictures of the pre-hydrogel prepared for loading in the 3D printed PCL scaffolds; Figure S2: Reaction process of hydrogel synthesis involved in this study. This reaction involved the activation of carboxyl groups of alginate by EDC to form active ester groups, followed by replacement of EDC by NHS to improve the efficiency of amine reaction. Afterward, NHS in NHS activated carboxylic group of alginate was replaced by primary amine of gelatin. Finally, CaCl$_2$ addition led to the ionic interactions of α-L-guluronic acid (G-block) of alginate to form hydrogel; Figure S3: STL image of gyroid-shaped three-dimensional cylindrical scaffold (a), side (b), and top (c) view of 3D printed gyroid scaffold of PCL, and SEM images of the scaffold to show the porous morphology (d); Figure S4: Standard curve, plotted between amounts of dissolved hydrogel and measured optical density at 630 nm; Figure S5: Representative SEM images of apatite grown on hydrogel (in PCL/hydrogel scaffold) after 12 days of immersion in SBF. The crack generated due to drying-induced strain was used to estimate the apatite layer thickness; Figure S6: Representative bright field (a) and fluorescence with 488 excitation laser (b) confocal microscopy images of hydrogel-loaded PCL scaffold without pre-stained cells to study the auto fluorescence of hydrogel, nano-hydroxyapatite and PCL scaffold. Results confirmed no auto fluorescence from hydrogel, nano-hydroxyapatite and PCL scaffold.

**Acknowledgments:** Binata Joddar acknowledges the NIH BUILD Pilot fund 8UL1GM118970-02 and NIH 1SC2HL134642-01for funding support. The authors acknowledge the use of core facilities at the Department of Metallurgical, Materials and Biomedical Engineering, University of Texas at El Paso for all experiments. The authors acknowledge the use of the Core Facility at Border Biomedical Research Consortium at UTEP supported by NIH-NIMHD-RCMI Grant No. 2G12MD007592.

**Author Contributions:** Ivan Hernandez has completed this study and data analysis as a part of his master thesis with mentoring and technical assistance from Alok Kumar. Alok Kumar conceived this idea for the work reported. Alok Kumar and Binata Joddar wrote and edited the manuscript. Binata Joddar supervised the whole project.

**Conflicts of Interest:** The authors declare no conflict of interest.

### References

1.  Lichte, P.; Pape, H.C.; Pufe, T.; Kobbe, P.; Fischer, H. Scaffolds for bone healing: Concepts, materials and evidence. *Injury* **2011**, *42*, 569–573. [CrossRef] [PubMed]
2.  Brydone, D. Bone grafting, orthopaedic biomaterials and the clinical need for bone engineering. *J. Eng. Med.* **2010**, *224*, 1329–1343. [CrossRef] [PubMed]
3.  Fabrizio, M.; Lorenzo, N.; Diana, C.; Massimo, I. New biomaterials for bone regeneration. *Clin. Cases Miner. Bone Metab.* **2011**, *8*, 21–24.
4.  Wang, Z.; Clark, C.C.; Brighton, C.T. Up-regulation of bone morphogenetic proteins in cultured murine bone cells with use of specific electric fields. *J. Bone Jt. Surg.* **2006**, *88*, 1053–1065. [CrossRef]

5. Dupont, K.M.; Sharma, K.; Stevens, H.Y.; Boerckel, J.D.; García, A.J.; Guldberg, R.E. Human stem cell delivery for treatment of large segmental bone defects. *Proc. Natl. Acad. Sci. USA* **2010**, *107*, 3305–3310. [CrossRef] [PubMed]

6. Chen, Y.; Xu, J.; Huang, Z.; Yu, M.; Zhang, Y.; Chen, H.; Ma, Z.; Liao, H.; Hu, J. An innovative approach for enhancing bone defect healing using plga scaffolds seeded with extracorporeal-shock-wave-treated bone marrow mesenchymal stem cells (BMSCS). *Sci. Rep.* **2017**, *7*, 44130. [CrossRef] [PubMed]

7. Gugala, Z.; Gogolewski, S. Healing of critical-size segmental bone defects in the sheep tibiae using bioresorbable polylactide membranes. *Injury* **2002**, *33*, 71–76. [CrossRef]

8. Yassin, M.A.; Leknes, K.N.; Pedersen, T.O.; Xing, Z.; Sun, Y.; Lie, S.A.; Finne-Wistrand, A.; Mustafa, K. Cell seeding density is a critical determinant for copolymer scaffolds-induced bone regeneration. *J. Biomed. Mater. Res. A* **2015**, *103*, 3649–3658. [CrossRef] [PubMed]

9. Luo, F.; Hou, T.-Y.; Zhang, Z.-H.; Xie, Z.; Wu, X.-H.; Xu, J.-Z. Effects of initial cell density and hydrodynamic culture on osteogenic activity of tissue-engineered bone grafts. *PLoS ONE* **2013**, *8*, e53697. [CrossRef] [PubMed]

10. Zhao, L.; Weir, M.D.; Xu, H.H.K. An injectable calcium phosphate-alginate hydrogel-umbilical cord mesenchymal stem cell paste for bone tissue engineering. *Biomaterials* **2010**, *31*, 6502–6510. [CrossRef]

11. Kumar, A.; Mandal, S.; Barui, S.; Vasireddi, R.; Gbureck, U.; Gelinsky, M.; Basu, B. Low temperature additive manufacturing of three dimensional scaffolds for bone-tissue engineering applications: Processing related challenges and property assessment. *Mater. Sci. Eng. R. Rep.* **2016**, *103*, 1–39. [CrossRef]

12. Mercado-Pagán, Á.E.; Stahl, A.M.; Shanjani, Y.; Yang, Y. Vascularization in bone tissue engineering constructs. *Ann. Biomed. Eng.* **2015**, *43*, 718–729. [CrossRef]

13. Kuijpers, A.J.; Engbers, G.H.M.; Krijgsveld, J.; Zaat, S.A.J.; Dankert, J.; Feijen, J. Cross-Linking and characterisation of gelatin matrices for biomedical applications. *J. Biomater. Sci. Polym. Ed.* **2000**, *11*, 225–243. [CrossRef] [PubMed]

14. Kuijpers, A.J.; Engbers, G.H.M.; Feijen, J.; de Smedt, S.C.; Meyvis, T.K.L.; Demeester, J.; Krijgsveld, J.; Zaat, S.A.J.; Dankert, J. Characterization of the network structure of carbodiimide cross-linked gelatin gels. *Macromolecules* **1999**, *32*, 3325–3333. [CrossRef]

15. Kokubo, T. *Bioceramics and their Clinical Applications*; Woodhead Publishing Limited: Cambridge, UK, 2008.

16. Ku, S.H.; Ryu, J.; Hong, S.K.; Lee, H.; Park, C.B. General functionalization route for cell adhesion on non-wetting surfaces. *Biomaterials* **2010**, *31*, 2535–2541. [CrossRef] [PubMed]

17. El-Sherbiny, I.M.; Yacoub, M.H. Hydrogel scaffolds for tissue engineering: Progress and challenges. *Glob. Cardiol. Sci. Pract.* **2013**, 38. [CrossRef]

18. Holy, C.E.; Shoichet, M.S.; Davies, J.E. Engineering three-dimensional bone tissue in vitro using biodegradable scaffolds: Investigating initial cell-seeding density and culture period. *J. Biomed. Mater. Res.* **2000**, *51*, 376–382. [CrossRef]

19. Tutak, W.; Kaufman, G.; Gelven, G.; Markle, C.; Maczka, C. Uniform, fast, high concentration delivery of bone marrow stromal cells and gingival fibroblasts by gas-brushing. *Biomed. Phys. Eng. Exp.* **2016**, *2*, 35007. [CrossRef]

20. Wong, R.S.H.; Ashton, M.; Dodou, K. Effect of crosslinking agent concentration on the properties of unmedicated hydrogels. *Pharmaceutics* **2015**, *7*, 305–319. [CrossRef] [PubMed]

21. Kuo, C.K.; Ma, P.X. Ionically crosslinked alginate hydrogels as scaffolds for tissue engineering: Part 1. Structure, gelation rate and mechanical properties. *Biomaterials* **2001**, *22*, 511–521. [CrossRef]

22. Hong, Y.; Huber, A.; Takanari, K.; Amoroso, N.J.; Hashizume, R.; Badylak, S.F.; Wagner, W.R. Mechanical properties and in vivo behavior of a biodegradable synthetic polymer microfiber–extracellular matrix hydrogel biohybrid scaffold. *Biomaterials* **2011**, *32*, 3387–3394. [CrossRef] [PubMed]

23. Bose, S.; Roy, M.; Bandyopadhyay, A. Recent advances in bone tissue engineering scaffolds. *Trends Biotechnol.* **2012**, *30*, 546–554. [CrossRef] [PubMed]

24. Burastero, G.; Scarfi, S.; Ferraris, C.; Fresia, C.; Sessarego, N.; Fruscione, F.; Monetti, F.; Scarfò, F.; Schupbach, P.; Podestà, M. The association of human mesenchymal stem cells with bmp-7 improves bone regeneration of critical-size segmental bone defects in athymic rats. *Bone* **2010**, *47*, 117–126. [CrossRef]

25. Murugan, R.; Ramakrishna, S. Development of nanocomposites for bone grafting. *Compos. Sci. Technol.* **2005**, *65*, 2385–2406. [CrossRef]

26. Jarcho, M. Calcium phosphate ceramics as hard tissue prosthetics. *Clin. Orthop. Rela. Res.* **1981**, *157*, 259–278. [CrossRef]

27. Oonishi, H.; Yamamoto, M.; Ishimaru, H.; Tsuji, E.; Kushitani, S.; Aono, M.; Ukon, Y. The effect of hydroxyapatite coating on bone growth into porous titanium alloy implants. *Bone Jt. J.* **1989**, *71*, 213–216.

28. Kumar, A.; Dhara, S.; Biswas, K.; Basu, B. In vitro bioactivity and cytocompatibility properties of spark plasma sintered Ha-ti composites. *J. Biomed. Mater. Res. B Appl. Biomater.* **2013**, *101B*, 223–236. [CrossRef] [PubMed]

29. Kumar, A.; Webster, T.J.; Biswas, K.; Basu, B. Flow cytometry analysis of human fetal osteoblast fate processes on spark plasma sintered hydroxyapatite–titanium biocomposites. *J. Biomed. Mater. Res. A* **2013**, *101*, 2925–2938. [CrossRef] [PubMed]

30. Kumar, A.; Nune, K.C.; Basu, B.; Misra, R.D.K. Mechanistic contribution of electroconductive hydroxyapatite–titanium disilicide composite on the alignment and proliferation of cells. *J. Biomater. Appl.* **2016**, *30*, 1505–1516. [CrossRef]

31. Agulhon, P.; Markova, V.; Robitzer, M.; Quignard, F.O.; Mineva, T. Structure of alginate gels: Interaction of diuronate units with divalent cations from density functional calculations. *Biomacromolecules* **2012**, *13*, 1899–1907. [CrossRef] [PubMed]

32. Rowley, J.A.; Madlambayan, G.; Mooney, D.J. Alginate hydrogels as synthetic extracellular matrix materials. *Biomaterials* **1999**, *20*, 45–53. [CrossRef]

33. Wu, S.-C.; Chang, W.-H.; Dong, G.-C.; Chen, K.-Y.; Chen, Y.-S.; Yao, C.-H. Cell adhesion and proliferation enhancement by gelatin nanofiber scaffolds. *J. Bioact. Compat. Polym.* **2011**, *26*, 565–577. [CrossRef]

34. Rammohan, A.V.; Lee, T.; Tan, V.B.C. A novel morphological model of trabecular bone based on the gyroid. *Int. J. Appl. Mech.* **2015**, *7*, 1550048. [CrossRef]

35. Yan, C.; Hao, L.; Hussein, A.; Young, P. Ti–6al–4v triply periodic minimal surface structures for bone implants fabricated via selective laser melting. *J. Mech. Behav. Biomed. Mater.* **2015**, *51*, 61–73. [CrossRef] [PubMed]

36. Yánez, A.; Herrera, A.; Martel, O.; Monopoli, D.; Afonso, H. Compressive behaviour of gyroid lattice structures for human cancellous bone implant applications. *Mater. Sci. Eng. C* **2016**, *68*, 445–448. [CrossRef] [PubMed]

37. Wang, M.O.; Vorwald, C.E.; Dreher, M.L.; Mott, E.J.; Cheng, M.H.; Cinar, A.; Mehdizadeh, H.; Somo, S.; Dean, D.; Brey, E.M. Evaluating 3D-printed biomaterials as scaffolds for vascularized bone tissue engineering. *Adv. Mater.* **2015**, *27*, 138–144. [CrossRef] [PubMed]

38. Mohammadi, H.; Hafezi, M.; Nezafati, N.; Heasarki, S.; Nadernezhad, A.; Ghazanfari, S.M.H.; Sepantafar, M. Bioinorganics in bioactive calcium silicate ceramics for bone tissue repair: Bioactivity and biological properties. *J. Ceram. Sci. Technol.* **2014**, *5*, 1–12.

39. Ichibouji, T.; Miyazaki, T.; Ishida, E.; Sugino, A.; Ohtsuki, C. Apatite mineralization abilities and mechanical properties of covalently cross-linked pectin hydrogels. *Mater. Sci. Eng. C* **2009**, *29*, 1765–1769. [CrossRef]

40. Heo, S.J.; Kim, S.E.; Wei, J.; Hyun, Y.T.; Yun, H.S.; Kim, D.H.; Shin, J.W.; Shin, J.W. Fabrication and characterization of novel nano-and micro-ha/pcl composite scaffolds using a modified rapid prototyping process. *J. Biomed. Mater. Res. A* **2009**, *89*, 108–116. [CrossRef] [PubMed]

41. Eglin, D.; Maalheem, S.; Livage, J.; Coradin, T. In vitro apatite forming ability of type i collagen hydrogels containing bioactive glass and silica sol-gel particles. *J. Mater. Sci. Mater. Med.* **2006**, *17*, 161–167. [CrossRef] [PubMed]

42. Hyun-Min, K.I.M.; Miyaji, F.; Kokubo, T.; Nakamura, T. Apatite-forming ability of alkali-treated ti metal in body environment. *J. Ceram. Soc. Jpn.* **1997**, *105*, 111–116.

43. Müller, P.; Bulnheim, U.; Diener, A.; Lüthen, F.; Teller, M.; Klinkenberg, E.D.; Neumann, H.G.; Nebe, B.; Liebold, A.; Steinhoff, G. Calcium phosphate surfaces promote osteogenic differentiation of mesenchymal stem cells. *J. Cell. Mol. Med.* **2008**, *12*, 281–291. [CrossRef] [PubMed]

44. Zhao, F.; Grayson, W.L.; Ma, T.; Bunnell, B.; Lu, W.W. Effects of hydroxyapatite in 3-D chitosan–gelatin polymer network on human mesenchymal stem cell construct development. *Biomaterials* **2006**, *27*, 1859–1867. [CrossRef] [PubMed]

45. Moreau, J.L.; Xu, H.H.K. Mesenchymal stem cell proliferation and differentiation on an injectable calcium phosphate–chitosan composite scaffold. *Biomaterials* **2009**, *30*, 2675–2682. [CrossRef] [PubMed]

46. Tagai, H.; Aoki, H. Preparation of Synthetic Hydroxyapatite and Sintering of Apatite Ceramics. In *Mechanical Properties of Biomaterials, Proceeding of Third Conference on Materials for Use in Medicine and Biology: Mechanical Properties of Biomaterials, Keele University, Keele, UK, 13–15 September 1978*; Hastings, G.W., Williams, D.F., Eds.; Wiley: Chichester, UK, 1980.

47. Kumar, A.; Biswas, K.; Basu, B. On the toughness enhancement in hydroxyapatite-based composites. *Acta Mater.* **2013**, *61*, 5198–5215. [CrossRef]

48. Oyane, A.; Kim, H.M.; Furuya, T.; Kokubo, T.; Miyazaki, T.; Nakamura, T. Preparation and assessment of revised simulated body fluids. *J. Biomed. Mater. Res. A* **2003**, *65*, 188–195. [CrossRef] [PubMed]

49. Wang, K.; Nune, K.C.; Misra, R.D.K. The functional response of alginate-gelatin-nanocrystalline cellulose injectable hydrogels toward delivery of cells and bioactive molecules. *Acta Biomater.* **2016**, *36*, 143–151. [CrossRef] [PubMed]

*gels*

MDPI

*Review*
# Polyampholyte Hydrogels in Biomedical Applications

Stephanie L. Haag and Matthew T. Bernards *

Department of Chemical & Materials Engineering, University of Idaho, Moscow, ID 83843, USA;
haag4885@vandals.uidaho.edu
* Correspondence: mbernards@uidaho.edu; Tel.: +1-208-885-2150

Received: 21 September 2017; Accepted: 3 November 2017; Published: 4 November 2017

**Abstract:** Polyampholytes are a class of polymers made up of positively and negatively charged monomer subunits. Polyampholytes offer a unique tunable set of properties driven by the interactions between the charged monomer subunits. Some tunable properties of polyampholytes include mechanical properties, nonfouling characteristics, swelling due to changes in pH or salt concentration, and drug delivery capability. These characteristics lend themselves to multiple biomedical applications, and this review paper will summarize applications of polyampholyte polymers demonstrated over the last five years in tissue engineering, cryopreservation and drug delivery.

**Keywords:** polyampholyte hydrogels; nonfouling; multi-functional

## 1. Introduction

A significant amount of research is being done with polyampholyte polymers in the biomedical community. Polyampholytes are polymeric systems comprised of both positively and negatively charged monomer subunits. Through the selection of monomers, one can build a polyampholyte with desired properties, tuned to specific biomedical applications. Our previous work evaluated much of the relevant literature prior to 2013 [1,2], so this paper is focused on advances over the past five years. We will first give a brief review of general polyampholyte characteristics with references to more thorough summaries, a discussion of the tunability of these systems, and an evaluation of recent findings using polyampholytes in tissue engineering, cryopreservation applications, and drug delivery.

## 2. General Polyampholyte Characteristics

A detailed explanation of the synthesis and properties of polyampholytes is beyond the scope of this paper and will not be provided here [3–6]. However, we will give a brief overview of the general characteristics that make polyampholytes attractive for biomedical applications. As mentioned above, polyampholytes contain both anionic and cationic functional groups. The strengths of these functional groups are often divided into four categories, that include both weak anionic and cationic groups, weak anionic and strong cationic groups, strong anionic and weak cationic groups, and lastly, both strong anionic and cationic groups. Table 1 shows the most commonly used monomers based on a survey of the recent literature. It should be noted that Table 1 is focused on summarizing synthetic organic monomer subunits. There is also a range of literature focused on naturally occurring materials that have been modified to include charged functional groups, like chitosan [7,8].

Table 1. Common Monomers Used in Polyampholyte Hydrogels.

| Chemical Name | Acronym | Monomer Formula | Strength of Functional Group |
|---|---|---|---|
| Acrylamide | AM | $CH_2=CHCONH_2$ | Weak cation |
| N-[3-(Dimethylamino)propyl] acrylamide | DMAPAA | $CH_2=CHCONH(CH_2)_3N(CH_3)_2$ | Weak cation |
| 2-(Dimethylamino)ethyl methacrylate | DMAEM | $CH_2=C(CH_3)COOCH_2CH_2N(CH_3)_2$ | Weak cation |
| 2-(Diethylamino)ethyl methacrylate | DEAEM | $CH_2=C(CH_3)CO_2CH_2CH_2N(C_2H_5)_2$ | Weak cation |
| [2-(Methacryloyloxy)ethyl] trimethylammonium chloride | TM | $CH_2=C(CH_3)CO_2CH_2CH_2N(CH_3)_3Cl$ | Strong cation |
| 2-(Acryloyloxy ethyl)trimethyl ammonium chloride | TMA | $CH_2=CHCO_2CH_2CH_2N(CH_3)_3Cl$ | Strong cation |
| [3-(Methacryloylamino)propyl] trimethylammonium chloride | MAPTAC | $CH_2=C(CH_3)CONH(CH_2)_3N(CH_3)_3Cl$ | Strong cation |
| 2-Carboxyethyl acrylate | CAA | $CH_2=CHCO_2(CH_2)_2CO_2H$ | Weak anion |
| Methacrylic acid | MAA | $CH_2=C(CH_3)COOH$ | Weak anion |
| Acrylic acid | AA | $CH_2=CHCOOH$ | Weak anion |
| Carboxylated poly-L-lysine | COOH-PLL | $NH_2(CH_2)_4CHNH_2COOH$ | Weak anion |
| 3-Sulfopropyl methacrylate potassium salt | SA | $H_2C=C(CH_3)CO_2(CH_2)_3SO_3K$ | Strong anion |
| 2-Sulfoethyl methacrylate | SE | $H_2C=C(CH_3)CO_2(CH_2)_2SO_3H$ | Strong anion |

Based on the selection of the underlying functional groups, polyampholytes have a tunable isoelectric point (IEP). The IEP occurs at the pH level when a polyampholyte is overall neutrally charged. The IEP is also the state at which a polyampholyte will have the most compact conformation, due to electrostatic attractions between the balanced, oppositely charged functional groups. As pH increases or decreases from the IEP, the overall charge of the polyampholyte will move further from neutral, causing electrostatic repulsive forces between like-charged regions, to increase and expand the polyampholyte. Similarly, when salt ions are present, the ions disrupt the electrostatic interactions between oppositely charged regions of the subunits. This also causes the polyampholyte to swell, as depicted schematically in Figure 1 [1]. The extent of swelling from pH or salt is ultimately dependent on the composition and architecture of the polymer [1,6]. However, manipulation of these unique electrostatic interactions and system responses has spurred investigation into using these materials in biomedical applications, as detailed throughout the rest of this review.

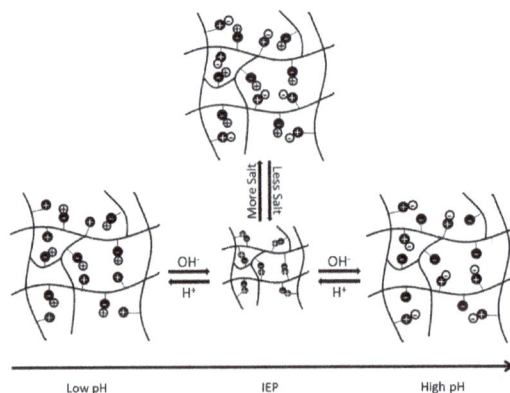

Figure 1. Schematic representation of the impact that changes in pH and salt concentrations have on electrostatic interactions within a polyampholyte hydrogel. This figure is reprinted from Ref. [1] with permission. Copyright 2013, Wiley Periodicals, Inc.

Another important general feature of overall charge neutral polyampholyte polymers is their natural nonfouling properties. It has been widely demonstrated [9–12] and reviewed previously [1,2] that this native resistance to nonspecific protein adsorption is the result of the formation of a strong hydration layer due to interactions between the naturally occurring dipole distribution in water and the charged regions of the underlying polyampholyte substrate. This is important because it is believed that this nonfouling property will lead to a reduced foreign body response in the in vivo environment, as seen with related zwitterionic systems. For example, in a previously reviewed paper, it was shown that polymer brushes composed of equimolar mixtures of 2-(acryl-oyloxy)ethyl trimethylammonium chloride (TMA) and 2-carboxyethyl acrylate (CAA) had only 4.3 ± 1.7 ng/cm$^2$ of nonspecific protein adsorption from 100% fetal bovine serum [13]. Furthermore, as demonstrated throughout the remainder of this review, pH changes can be used to modify the net neutral charge of polyampholyte systems, adding in a responsive component to the utilization of these polymers in biomedical applications.

## 3. Mechanical Properties

The composition dependent tunability of polyampholyte systems also provides a unique approach for easily controlling the mechanical properties of the biomaterial. To facilitate better tissue regeneration and integration, it is important for an implanted biomaterial to mimic the native properties of the tissues it is supplanting [14,15]. There is, of course, great variability in the mechanical properties of tissues, as properties range from soft and flexible (skin) to strong, with the ability to absorb impact forces (bone). In addition, biomaterials must also have a high water content, to maintain their biocompatibility, and the ability for cells to penetrate into the material.

Our group demonstrated the easy tunability of polyampholyte hydrogels utilizing various ratios of monomers in three component hydrogels consisting of positively charged 2-(Acryl-oyloxy)ethyl trimethylammonium chloride (TMA), and varying mixtures of negatively charged 2-carboxyethyl acrylate (CAA) and 3-sulfopropyl methacrylate (SA) monomers [16]. Furthermore, the crosslinker density was also used as a mechanism for further tuning the mechanical properties. It was demonstrated that both the density of the crosslinker, as well as the ratio of monomers in the hydrogel, altered the fracture strength and Young's Modulus. At crosslinker densities of 1× and 2× (1:0.076 and 1:0.152 monomer/crosslinker ratios), the mechanical properties were dependent upon the exact combination of monomer subunits, while at a 4× crosslinker density, the crosslinker became the controlling factor. However, this study clearly demonstrated the easily tuned mechanical properties of polyampholyte systems with low crosslinker densities. In a similar fashion, Jian and Matsumura were able to controllably tune the mechanical properties of their nanocomposite hydrogel designed with carboxylated poly-L-lysine (COOH-PLL), and synthetic clay laponite XLG, by changing the laponite concentration (composition dependence) or the density of the polyethylene glycol with N-hydroxy succinimide ester (PEG–NHS) crosslinker [17]. Changing the crosslinker density or monomer concentration are also common tuning mechanisms for mechanical properties [18–22].

A great deal of both theoretical and experimental study has been conducted to better understand the fracture mechanisms of polyampholyte gels, for use in guiding the design of stronger or more tunable systems [23]. Above a critical loading stress, moderately chemically crosslinked hydrogels resisted creep flow, while physically crosslinked and lightly chemically crosslinked hydrogels experience creep rupture. However, at large stresses, creep behavior indicated that both physically and chemically crosslinked hydrogels undergo bond breaking mechanisms. These results confirm that chemical bonds are stronger than physical bonds, therefore, chemically crosslinked systems show an improvement over systems with only ionic bonds [24]. However, the incorporation of physical crosslinks has positively influenced fracture behavior of viscoelastic hydrogels through reduced deformation rate [25] and crack blunting [26].

Due to the beneficial features of both chemical and physical crosslinks, recent studies have approached the development of mechanically strong hydrogels by combining the two mechanisms in an approach referred to as the sacrificial bond principle [14,20,27–32]. The sacrificial bond principle is based on the formation of a highly stretchable base matrix, with a high density of brittle sacrificial bonds that are weaker than the base matrix. During stress, the brittle bonds break before the stretchable base matrix, leading to improved mechanical performance. Figure 2 shows a schematic of possible fracture processes with and without sacrificial bonds present [27].

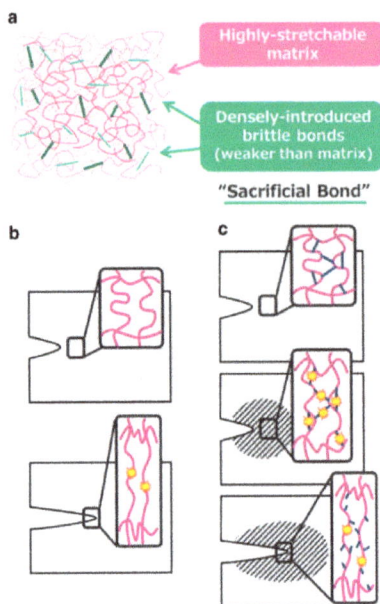

**Figure 2.** (**a**) General structure of a tough gel based on the sacrificial bond principle consisting of a highly stretchable matrix with a high density of brittle bonds; (**b**) Possible fracture processes of a single network gel; (**c**) Possible fracture processes of a sacrificial bond gel. The brittle bonds are widely ruptured prior to the macroscopic crack propagation around the crack tip (shadowed zone). This figure is reprinted from Ref. [27] with permission. Copyright 2017, the Society of Polymer Science, Japan.

These sacrificial bonds can be covalent bonds, hydrogen bonds, ionic bonds, or hydrophobic interactions, depending on the polymer matrix. They can also be incorporated into the base matrix with multiple approaches, including double network gels, ionically linked gels, metal ion chelation, and composite gels [27,33]. The resulting hydrogels from all of these approaches show great mechanical strength, energy dissipation, and force dispersion, to slow down fracture and crack propagation [20]. In just one representative example, the use of a double network hydrogel composed of poly(2-acrylamido-2-methylpropanesulfonic acid) and poly(acrylamide) was shown to improve the compressive fracture stress from 0.4–0.8 to 17.2 MPa [34].

With polyampholyte double network hydrogels showing irreversible deformation, however, efforts started on the use of other types of bonds that could be reversible and self-healing. Some work has been done using electrostatic interactions and hydrophobic interactions. The mechanical properties are extremely dependent on pH, as the interactions that hold the structure together can be weak or strong, depending on the charged state of the monomers. Furthermore, the material will also swell and collapse with changes in pH [28,29]. One study added partially quaternized poly(4-vinylpyridine) into an elastic hydrogel, thereby introducing electrostatic, hydrophobic, and hydrogen bonding interactions to better dissipate energy. This resulted in an increase in fracture energy from 44 to 1000 J/m$^2$ [30].

In polyampholytes, it is common to take advantage of the electrostatic interactions as a secondary sacrificial bond, to toughen materials via the presence of oppositely charged functional groups distributed throughout the system. Strong electrostatic interactions act as permanent crosslinks, and weaker interactions reversibly break and reform, which dissipates energy and toughens the gels [14,32]. These bonds can also occur via both inter- and intra-chain interactions. Polyampholytes and polyion complex hydrogels (PIC) both contain oppositely charged functional groups and have potential as tough, self-healing gels. PICs are formed from electrostatic interactions between oppositely charged polyelectrolye polymers upon mixing. Polyampholytes form the toughest hydrogels around zero net charge, where PIC systems can form tough gels at weakly off-balanced charge compositions. PICs are typically tougher than polyampholytes when they have the same monomer compositions, due to the fact that PIC hydrogels form at lower concentrations than polyampholytes [31].

Additional approaches have also been used to improve the mechanical properties of hydrogels based on ionic bonding. In one example, the removal of co-ions prior to gelation was shown to facilitate improved ionic bond formation [35]. In another study, Cui et al. developed a method referred to as pre-stretching, where hydrogels are prepared and then stretched. This stretching helps align the chains parallel to each other, as opposed to the original random alignment. When the chains are parallel, stronger ionic bonds form, which in turn strengthens the overall polyampholyte hydrogel [36]. Fang et al. explored a similar approach to attain a tough and stretchable hydrogel by altering the structure of the material [37]. Starting with a protein-based hydrogel, they forced the unfolding of the globular domains. The subsequent collapse and aggregation of the unfolded material allows for physical intertwining and linking through electrostatic interactions. The resulting hydrogels have the unusual properties of a negative swelling ratio, high stretchability, and toughness.

Byette et al. took inspiration from the mechanisms used by mussels to attach to wet surfaces as an approach to toughen polyampholyte materials [38]. Mussels use byssus, a protein-based material, to secure themselves to solid surfaces. Byssus shows a self-healing ability combined with strength, partially due to metal ions forming sacrificial bonds with the amino acid subunits. Byette et al. created a hydrogel from byssus protein hydrolysate, and treated it with $Ca^{2+}$ or $Fe^{3+}$. The films with $Fe^{3+}$ showed the greatest increase in strength and toughness. A similar approach was used by Huang et al., who made a semi-interpenetrating polymer network composed of carboxymethyl chitosan (CMCH), acrylamide, and maleic acid with carboxylic–$Fe^{3+}$ interactions serving as ionic sacrificial bonds [39]. By changing the ratio of maleic acid and the concentration of $Fe^{3+}$, the best hydrogels showed a tensile stress of 1.44 MPa. Additionally, the CMCH provided the gels with antibacterial characteristics against *Staphylococcus aureus* and Gram negative *Escherichia coli*.

## 4. Tissue Engineering Applications

Polyampholyte hydrogels are an attractive option for tissue engineering, due to the general characteristics described above. In addition to their tunable, responsive, and nonfouling properties, they also have a high moisture holding capacity, which is generally associated with biocompatibility. Our group has demonstrated multifunctional polyampholyte hydrogels for tissue engineering using TMA and CAA monomer subunits [40,41]. These gels show excellent resistance to nonspecific protein adsorption, including negatively charged fibrinogen (FBG) and positively charged lysozyme (LYZ), and they prevent the short-term adhesion of MC3T3-E1 cells [41]. The elimination of nonspecific cell adhesion is intended to reduce the occurrence of the foreign body response in the in vivo environment, but it is not desirable for facilitating tissue regeneration through the implanted scaffold. However, the multifunctional capabilities of the polyampholyte hydrogel platform demonstrated in this work provides an easy mechanism for incorporating cell adhesive biological cues. The pH responsive nature of the CAA monomer can be taken advantage of with the use of *N*-(3-dimethylaminopropyl)-*N*′-ethylcarbodiimide hydrochloride/*N*-hydroxysuccinimide (EDC/NHS) bioconjugation chemistry to covalently attach bioactive signaling molecules. This was used to attach FBG, which subsequently facilitated MC3T3-E1 cell adhesion to the hydrogel, as demonstrated in Figure 3 [41]. Furthermore, the background hydrogel (locations without conjugated FBG) was tested, and verified that it retained the native nonfouling properties away from the conjugated proteins, upon return to neutral pH [41]. It is believed that the incorporation of tissue specific biological cues will facilitate targeted cell adhesion and interrogation interactions. This multifunctional capability is not limited to just TMA/CAA polyampholyte hydrogels, either. Three component polymers, using equimolar combinations of positively charged TMA and varying combinations of negatively charged CAA and SA monomers, have also shown the same nonfouling properties and pH dependent protein conjugation capabilities, regardless of the underlying charge balanced composition [16]. This combination of nonfouling properties, protein conjugation capability, and tunable cell adhesion suggests polyampholyte hydrogels have excellent potential for applications as tissue engineering scaffolds.

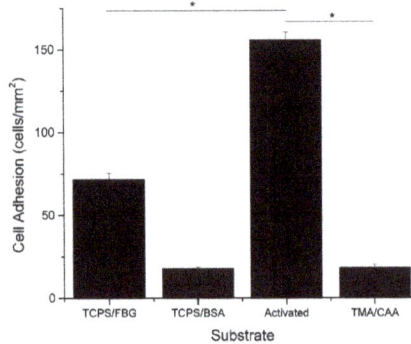

**Figure 3.** Average number of MC3T3-E1 cells (cells/mm$^2$) that adhered to tissue culture polystyrene (TCPS) and TMA/CAA hydrogels with or without adsorbed or conjugated proteins. * Represents a statistically significant difference between the surfaces being compared ($p < 0.05$). This figure is reprinted from Ref. [41] with permission. Copyright 2013, American Chemical Society.

Advances in the application of polyampholyte hydrogels for tissue engineering are not limited to our efforts. For example, Jian and Matsumura developed a nanocomposite hydrogel using COOH–PLL and synthetic clay laponite XLG, that showed promise as a tissue engineering scaffold due to its controlled release profiles, good mechanical properties, and cell adhesion capability [17]. These gels were cytocompatible, and had adjustable degradation properties. Furthermore, cell adhesion was tunable by controlling the hydrogel formulation. When the polymer chains were covalently cross-linked with PEG-NHS, it hid some of the laponite surface and reduced cell adhesion. Alternatively, when the hydrogels were only physically crosslinked (no PEG–NHS), there was more exposed laponite surface area, leading to enhanced cell attachment.

## 5. Cryopreservation Applications

Another important aspect of tissue engineering is the preservation of cells over long-term scenarios. This is most generally done using cryopreservation in a liquid nitrogen cell freezer. In order to prevent cell death, a cryoprotective agent (CPA) is typically added to the cell solution prior to freezing. One of the most commonly used CPAs is dimethyl sulfoxide (DMSO), but it shows high cytotoxicity and needs to be removed quickly after thawing. DMSO has also been seen to influence the differentiation of many cell types. The need for a new and more effective CPA has driven research into the use of polyampholytes for cryopreservation.

Matsumura et al. demonstrated the use of COOH-PLL as a new polyampholyte CPA for human bone marrow derived mesenchymal stem cells (hBMSCs) [42]. They found that the polyampholyte CPA did not penetrate the cell wall, but instead provided protection by attaching to the membrane. When the ratio of carboxylation was within the range of 0.5–0.8, there was >90% cell viability upon seeding after being frozen for 24 months, with no significant differences compared to cells frozen in the presence of DMSO. The hBMSCs also showed better retention of their properties inherent before freezing, such as differentiation potential, as compared to samples with DMSO as the CPA. COOH-PLL was further tested as a CPA during fast and slow vitrification of two-dimensional cell constructs. Figure 4a,b below, show the cell viability directly after warming, and after one day of culture. It can be clearly seen that there are no significant differences in the cell viability immediately after thawing in the presence of COOH-PLL (denoted as P-VS), DMSO (denoted as DAP213), or no CPA (denoted as VS). After both one day of culture and over longer time periods, the proliferation curves (Figure 4c) show a distinct improvement when the cells were frozen with the polyampholyte CPA, as compared to either DMSO or no CPA. Through these studies, it was concluded that the use of a polyampholyte CPA significantly improved the viability of hBMSCs while maintaining differentiation capacity, making it promising for the long-term storage of tissue engineered constructs [42,43].

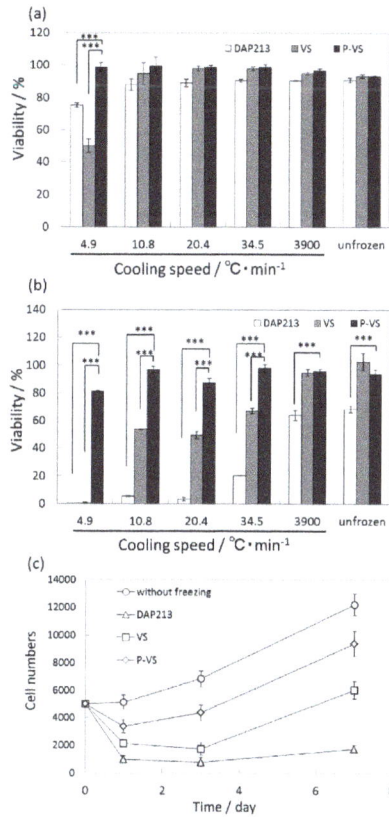

**Figure 4.** Quantitative viability results of mesenchymal stem cells (MSCs) after slow and fast vitrification with various VSs and different cooling speeds (**a**) immediately after warming and (**b**) after 1 day of culture. (**c**) Cell proliferation curves after slow vitrification at a cooling rate of 10.8 °C/min with various VSs (*** $p < 0.001$). This figure reprinted from Ref. [43] with permission. Copyright 2016, American Chemical Society.

Based on the positive results seen with COOH-PLL, other polyampholytes have also been investigated as CPAs. These studies were to both expand the formulation range of CPAs, as well as to better understand how polyampholytes protect the cell membrane during freezing. In one example, 2-(dimethylamino) ethyl methacrylate (DMAEMA) and methacrylic acid (MAA) were copolymerized in various ratios [44]. In addition, hydrophobic groups in the form of *N*-butyl methacrylate (Bu-MA) and *N*-octyl methacrylate (Oc-MA) were introduced into the polymer backbone at 2–10% mole percent of the total monomer amount. This range of polyampholyte chemistries were tested, and at an overall solution polymer concentration of 10%, with 5% consisting of Bu-MA or Oc-MA, significantly increased cell viability was seen following freezing. By testing this range of polyampholyte compositions, it was determined that the cryoprotective properties are strongly correlated with hydrophobicity. This approach has also been adapted to the closely related zwitterionic polymers 3-((3-acrylamidoproply)dimethylammonio)-propane-1-sulfonate and 2-((2-methacryloyloxy)ethyl)-dimethylamino acetate [45]. The cryoprotective capabilities of the zwitterionic species were compared to poly(MAA-DMAEMA), and they did not show comparable cell viability, providing further insight into the mechanism of preservation. Through these studies, it was concluded that the cryoprotective property results from strong interactions between the polyampholyte CPA and the cell membrane, which are greatly aided by limited hydrophobic interactions [44,45].

Cell sheets and constructs have added complexity for successful cryopreservation. A dextran-based polyampholyte hydrogel was developed to encapsulate cell constructs prior to cryopreservation, and it has shown promise for tissue engineering applications in preliminary studies [46]. Another variation on the use of COOH-PLL CPAs was explored by Jian and Matsumura. Cells were cryopreserved with 7.5–20% COOH-PLL solutions. After thawing, nanosilicates were injected, turning the solution into a thixotropic hydrogel. Cell viability was excellent, remaining >90% for all tested polyampholyte concentrations. This unique gel system was proposed for direct cell injection for site specific cell delivery and tissue repair, without the need to wash out the cryoprotective agent [47]. Furthermore, the thermoresponsiveness of this class of polyampholyte materials, and their demonstrated biocompatibility, make them promising for other biomaterial and drug delivery applications [48].

## 6. Drug Delivery Applications

Due to the naturally occurring responsive nature of polyampholyte polymers addressed earlier, they have gained increasing interest for drug delivery applications. The cryoprotective properties of some polyampholyte formulations, discussed above, have been taken a step further by Ahmed et al. as a novel approach to deliver proteins into cells [49]. Proteins were adsorbed on/into nanoparticles made from hydrophobically modified polyampholytes synthesized by the succinylation of ε-poly-L-lysine with dodecyl succinic anhydride and succinic anhydride. L929 cells were then frozen with the protein loaded nanoparticles as a CPA. The high affinity between the cell membrane and the hydrophobic subunits of the nanoparticles caused the protein-loaded nanoparticles to condense on the peripheral cell membrane during freezing. The adsorbed protein and nanoparticles were found to be internalized after thawing via endocytosis during culture, thereby delivering the protein payload. However, there was a critical concentration above which these nanoparticle delivery systems became cytotoxic. The Matsumura group also adapted this approach to polyampholyte-modified liposomes in additional protein delivery studies, demonstrating its adaptability for protein delivery in immunotherapy applications [50].

At the same time, much of the recent work in polyampholyte mediated drug delivery takes advantage of the pH responsive behavior of polyampholyte systems. For example, chitosan based polyampholytes have recently been shown to have potential in protein delivery applications, as they have exhibited the ability to adsorb and desorb bovine serum albumin (BSA) in a pH dependent manner [7,8]. However, a combination of design characteristics is required to optimize drug delivery that include biocompatibility, multifunctionality, and responsiveness to the microenvironment. Nanogels have been investigated for use as delivery systems, and have shown tremendous promise due to the ability to control drug release, provide the drug protection from degradation, and target specific tissues. Some of the loading and drug release methods include covalent conjugation, passive/diffusion based, or through environmental stimuli, such as pH [51].

Our group previously investigated the fundamental release characteristics of polyampholyte hydrogels composed of equimolar ratios of TMA and CAA using neutral caffeine, positively charged methylene blue, and negatively charged metanil yellow [52]. These species were selected as methylene blue and metanil yellow have nearly identical molecular weights, thereby eliminating this variable when comparing the release kinetics, while caffeine is approximately one half the size of the other species, to allow for a characterization of the influence of size. Hydrogels were synthesized in the presence of the drug analogues, and then the release characteristics were monitored as a function of crosslinker density, pH, and ionic concentration. The release of the smaller, neutral caffeine molecule was shown to be mediated by diffusion alone, although this release was tunable, based on environmental stimuli induced swelling of the polyampholyte hydrogels. Conversely, the release of the charged molecules was strongly dependent on electrostatic interactions throughout the system, which could be modified through the environmental cues of pH and ionic strength. Figure 5 shows a schematic of the relative drug release levels from the TMA/CAA hydrogel. Importantly, it was also demonstrated that following the release of the various drug molecules, it was verified that the TMA/CAA platforms retained their native nonfouling characteristics. Therefore, this platform shows great potential for long-term biomolecule delivery.

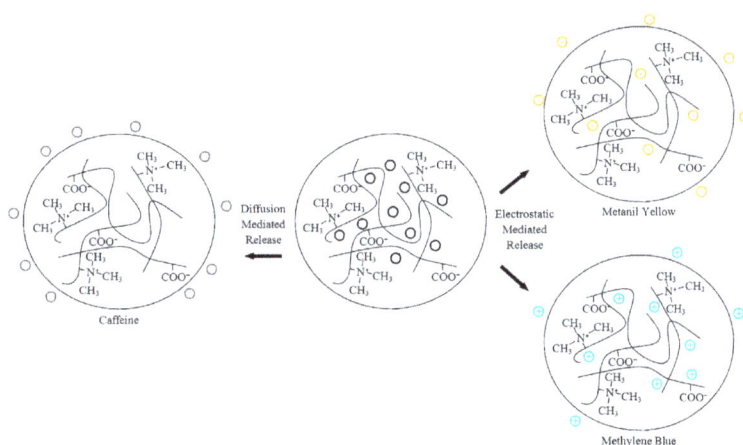

**Figure 5.** Schematic depicting the release of caffeine, metanil yellow and methylene blue from TMA/CAA gels. This figure is reprinted from Ref. [52] with permission. Copyright 2015, American Chemical Society.

Kudaibergenov et al. also used a variety of guest molecules to characterize the adsorption and release from a macroporous amphoteric cryogel composed of *N*,*N*-dimethylaminoethyl methacrylate and methacrylic acid with a *N*,*N*′-methylenebisacrylamide crosslinker [53]. The guest species tested included methylene blue, methyl orange, sodium dodecylbenzene sulfonate (SDBS), and lysozyme. Lysozyme and methylene blue were adsorbed at pH 9.5, and SDBS and methyl orange were adsorbed at pH 7.5. Similar to the work by Barcellona et al., the binding interactions between the cryogel and the guest molecules was driven by electrostatic forces. However, at the IEP of pH 7.1, the amphoteric cryogel allowed for the release of 93–98% of the adsorbed species. The conclusions drawn by both Barcellona et al. and Kudaibergenov et al. are also supported by simulation based studies that concluded that electrostatic interactions play the most significant role in mediating drug release from polyampholyte systems [54].

A variety of specific drug species have also been used to test drug delivery from assorted polyampholyte mediums. Mishra et al. used poly 3-[(methacryoylamino)propyl trimethylammonium chloride-*co*-methacrylic acid] (PMAPTACMAAc) copolymers with various concentrations of monomers and loaded indomethacin (IND) [55]. IND is a nonsteroidal anti-inflammatory drug that is used for the treatment of rheumatoid arthritis, ankylosing spondylitis, and osteoarthritis, to name a few. The hydrogel composition played a large role in the sustained release of IND, and PMAPTACMMAc-5 led to the highest percentage of IND release. This formulation released 75% of the entrapped IND within 8 h and 82% after 12 h. Other hydrogel formulations showed release percentages ranging from ~44% to 77% after 12 h. The release was primarily diffusion based, and it followed non-Fickian release kinetics. Although diffusion is often effective for drug delivery, a controlled release response can provide a more targeted delivery. Salicylic acid was used as a model drug in a polyampholyte composed of casein and poly($N$-isoproplyacrylamide), and the release was affected by temperature, pH, and crosslinker density [56]. This led Cao et al. to conclude this delivery vehicle was appropriate for orally administered drug delivery. Finally, Sankar et al. demonstrated the pH sensitive release of promethazine hydrochloride from polyampholyte hydrogels containing carbon nanotubes [57]. These nanotubes were incorporated into the hydrogel as an approach to reinforce the mechanical properties of this delivery system, without impacting the drug delivery capabilities.

Investigators have also begun incorporating polyampholytes into multicomponent systems to enhance performance or offer additional benefits. For example, Wang et al. examined a polyampholyte hydrogel release system based on pyromellitic diester diacid chloride (PDDC) combined with combinations of diethylenetriamine (DETA) and triazine [58]. This polyampholyte system showed a pH dependent release capability that overcame previous issues seen with related encapsulants formed with terephthaloyl chloride (TC) in place of PDDC. This new microcapsule formulation showed high loading capacity, and steady, controlled release at pH 7.4. It also demonstrated accelerated release at both pH 5 and pH 10, as shown in Figure 6. The release characteristics were also tunable by varying the ratio of DETA to triazine, indicating the ability to refine this microcapsule formulation for tunable release rate applications.

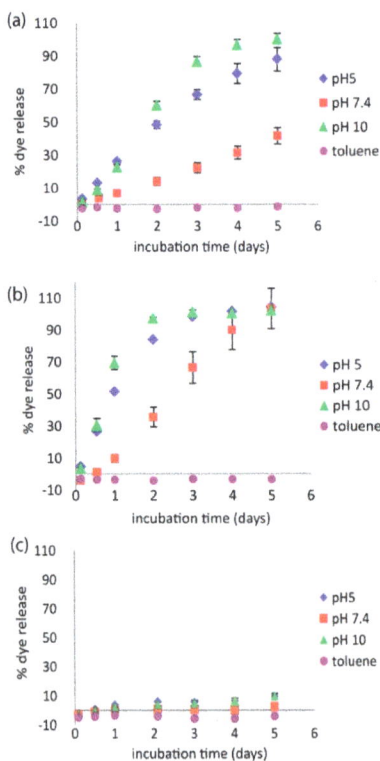

**Figure 6.** Release profiles of coumarin 1 dye under different solvent conditions for PDDC capsules with (**a**) 3:1 triazine/DETA and (**b**) 1:1 triazine/DETA. (**c**) Control experiments: 1:1 triazine/DETA with TC. This figure is reprinted from Ref. [58] with permission. Copyright 2017, American Chemical Society.

Others has also incorporated polyampholyte polymers into their drug delivery vehicles to add pH responsive release characteristics. For example, Schulze et al. saw potential in lamellar liquid crystalline systems, but the structure did not react to environmental stimuli such as pH. When the polyampholyte poly(*N*,*N'*-diallyl-*N*,*N'*-dimethyl-almaleamic carboxylate) (PalH) was integrated into a lamellar liquid crystalline system of sodium dodecyl sulfate, decanol, and water, it was found that release from the new structure could be tuned by varying the pH or temperature. This suggests it has promise as a new structural material for drug delivery systems [59]. In another example, papacetamol, an analgesic drug, was released from a polyampholyte hydrogel matrix composed of laponite, polyacrylamide and poly(3-acrylamidopropyl) trimethylammonium chloride. Drug release was tested as a function of environmental changes in pH and ionic strength, and in the presence of an electric field. Without an electric field, papacetamol was only released at pH 1.1, but with the application of an electric field, sustained drug release occurred at other pH values [60]. Finally, Ali et al. created a novel polymer containing residues of alendronic acid, that showed pH sensitive responses that were proposed to be used as a drug delivery system [61].

Asayama et al. also incorporated a polyampholyte polymer, carboxymethyl poly(1-vinylimidazole) (CM-PVIm), into an existing system. CM-PVIm was used to coat poly(ethylenimine)/DNA (PEI/DNA) complexes, to reduce nonspecific protein adsorption to this delivery platform. The results demonstrated this coating did not significantly reduce gene transfection or cell viability. Therefore, the authors concluded that CM-PVIm is an effective coating for improved circulation of gene therapy agents [62].

Another application of adding polyampholytes to drug delivery vehicles is based on their strong water holding capacity. Polyampholyte acrylic latexes were incorporated into drug tablet coatings to minimize the amount of water removed from the drugs during the tablet drying step [63]. During the optimization of this approach, Ladika et al. focused on finding a polymer solution with similar viscosity to the industry standard, that contained a much higher concentration of solids. Typical tablet coatings on the market today range from 4–10 wt % solids and the new polyampholyte acrylic latexes showed a range of 37–39 wt % solids. Three types of latexes were explored: weak acid/strong base latexes, strong acid/weak base latexes, and combinations of anionic and cationic latexes. Latex formulations for all three combinations were determined that had viscosities similar to current coating solutions, had higher solids composition, and were pH-tunable, to enable targeted delivery of active pharmaceutical ingredients.

## 7. Future Directions

Throughout this review, many exciting advancements applying polyampholyte hydrogels to biomedical applications were highlighted. However, despite this progress and the clearly demonstrated capabilities of polyampholytes, these materials have not yet been investigated in the in vivo environment in depth. This is the critical next step in the continued development of these materials, and our group is pursuing these efforts in the application of polyampholyte hydrogels for bone tissue engineering. Additionally, while the tunability and responsive properties of polyampholytes have been widely demonstrated, the ease of tuning polyampholyte materials for targeted applications of these capabilities must also be further pursued.

**Acknowledgments:** This research was supported in part by a grant from the Department of Defense through grant W81XWH-15-1-0664.

**Author Contributions:** The review paper was co-written by Stephanie L. Haag and Matthew T. Bernards.

**Conflicts of Interest:** The authors declare no conflict of interest.

## References

1.  Zurick, K.M.; Bernards, M. Recent Biomedical Advances with Polyampholyte Polymers. *J. Appl. Polym. Sci.* **2014**, *131*. [CrossRef]

2.    Bernards, M.; He, Y. Polyampholyte polymers as a versatile zwitterionic biomaterial platform. *J. Biomater. Sci.-Polym. Ed.* **2014**, *25*, 1479–1488. [CrossRef] [PubMed]
3.    Laschewsky, A. Structures and Synthesis of Zwitterionic Polymers. *Polymers* **2014**, *6*, 1544–1601. [CrossRef]
4.    Gao, M.; Gawel, K.; Stokke, B.T. Polyelectrolyte and antipolyelectrolyte effects in swelling of polyampholyte and polyzwitterionic charge balanced and charge offset hydrogels. *Eur. Polym. J.* **2014**, *53*, 65–74. [CrossRef]
5.    Kudaibergenov, S.E.; Nuraje, N.; Khutoryanskiy, V.V. Amphoteric nano-, micro-, and macrogels, membranes, and thin films. *Soft Matter* **2012**, *8*, 9302–9321. [CrossRef]
6.    Lowe, A.B.; McCormick, C.L. Synthesis and solution properties of zwitterionic polymers. *Chem. Rev.* **2002**, *102*, 4177–4189. [CrossRef] [PubMed]
7.    Kono, H.; Oeda, I.; Nakamura, T. The preparation, swelling characteristics, and albumin adsorption and release behaviors of a novel chitosan-based polyampholyte hydrogel. *React. Funct. Polym.* **2013**, *73*, 97–107. [CrossRef]
8.    Yilmaz, E.; Yalinca, Z.; Yahya, K.; Sirotina, U. pH responsive graft copolymers of chitosan. *Int. J. Biol. Macromol.* **2016**, *90*, 68–74. [CrossRef] [PubMed]
9.    Shen, X.; Yin, X.; Zhao, Y.; Chen, L. Antifouling enhancement of PVDF membrane tethered with polyampholyte hydrogel layers. *Polym. Eng. Sci.* **2015**, *55*, 1367–1373. [CrossRef]
10.   Shen, X.; Yin, X.B.; Zhao, Y.P.; Chen, L. Improved protein fouling resistance of PVDF membrane grafted with the polyampholyte layers. *Colloid Polym. Sci.* **2015**, *293*, 1205–1213. [CrossRef]
11.   Zhao, T.; Chen, K.M.; Gu, H.C. Investigations on the Interactions of Proteins with Polyampholyte-Coated Magnetite Nanoparticles. *J. Phys. Chem. B* **2013**, *117*, 14129–14135. [CrossRef] [PubMed]
12.   Peng, X.L.; Zhao, L.; Du, G.F.; Wei, X.; Guo, J.X.; Wang, X.Y.; Guo, G.S.; Pu, Q.S. Charge Tunable Zwitterionic Polyampholyte Layers Formed in Cyclic Olefin Copolymer Microchannels through Photochemical Graft Polymerization. *ACS Appl. Mater. Interfaces* **2013**, *5*, 1017–1023. [CrossRef] [PubMed]
13.   Tah, T.; Bernards, M.T. Nonfouling polyampholyte polymer brushes with protein conjugation capacity. *Colloids Surf. B Biointerfaces* **2012**, *93*, 195–201. [CrossRef] [PubMed]
14.   Sun, T.L.; Kurokawa, T.; Kuroda, S.; Bin Ihsan, A.; Akasaki, T.; Sato, K.; Haque, M.A.; Nakajima, T.; Gong, J.P. Physical hydrogels composed of polyampholytes demonstrate high toughness and viscoelasticity. *Nat. Mater.* **2013**, *12*, 932–937. [CrossRef] [PubMed]
15.   Huang, Y.W.; King, D.R.; Sun, T.L.; Nonoyama, T.; Kurokawa, T.; Nakajima, T.; Gong, J.P. Energy-Dissipative Matrices Enable Synergistic Toughening in Fiber Reinforced Soft Composites. *Adv. Funct. Mater.* **2017**, *27*. [CrossRef]
16.   Cao, S.; Barcellona, M.N.; Pfeiffer, F.; Bernards, M.T. Tunable multifunctional tissue engineering scaffolds composed of three-component polyampholyte polymers. *J. Appl. Polym. Sci.* **2016**, *133*. [CrossRef]
17.   Jain, M.; Matsumura, K. Polyampholyte- and nanosilicate-based soft bionanocomposites with tailorable mechanical and cell adhesion properties. *J. Biomed. Mater. Res. Part A* **2016**, *104*, 1379–1386. [CrossRef] [PubMed]
18.   Bin Ihsan, A.; Sun, T.L.; Kuroda, S.; Haque, M.A.; Kurokawa, T.; Nakajima, T.; Gong, J.P. A phase diagram of neutral polyampholyte—From solution to tough hydrogel. *J. Mater. Chem. B* **2013**, *1*, 4555–4562. [CrossRef]
19.   Luo, F.; Sun, T.L.; Nakajima, T.; Kurokawa, T.; Li, X.F.; Guo, H.L.; Huang, Y.W.; Zhang, H.J.; Gong, J.P. Tough polyion-complex hydrogels from soft to stiff controlled by monomer structure. *Polymer* **2017**, *116*, 487–497. [CrossRef]
20.   Wang, H.W.; Li, P.C.; Xu, K.; Tan, Y.; Lu, C.G.; Li, Y.L.; Liang, X.C.; Wang, P.X. Synthesis and characterization of multi-sensitive microgel-based polyampholyte hydrogels with high mechanical strength. *Colloid Polym. Sci.* **2016**, *294*, 367–380. [CrossRef]
21.   Li, G.; Zhang, G.P.; Sun, R.; Wong, C.P. Dually pH-responsive polyelectrolyte complex hydrogel composed of polyacrylic acid and poly (2-(dimthylamino) ethyl methacrylate). *Polymer* **2016**, *107*, 332–340. [CrossRef]
22.   Wang, L.; Wang, H.H.; Yu, H.C.; Luo, F.; Li, J.H.; Tan, H. Structure and properties of tough polyampholyte hydrogels: Effects of a methyl group in the cationic monomer. *RSC Adv.* **2016**, *6*, 114532–114540. [CrossRef]
23.   Long, R.; Hui, C.Y. Fracture toughness of hydrogels: Measurement and interpretation. *Soft Matter* **2016**, *12*, 8069–8086. [CrossRef] [PubMed]
24.   Karobi, S.N.; Sun, T.L.; Kurokawa, T.; Luo, F.; Nakajima, T.; Nonoyama, T.; Gong, J.P. Creep Behavior and Delayed Fracture of Tough Polyampholyte Hydrogels by Tensile Test. *Macromolecules* **2016**, *49*, 5630–5636. [CrossRef]

25. Sun, T.L.; Luo, F.; Hong, W.; Cui, K.P.; Huang, Y.W.; Zhang, H.J.; King, D.R.; Kurokawa, T.; Nakajima, T.; Gong, J.P. Bulk Energy Dissipation Mechanism for the Fracture of Tough and Self-Healing Hydrogels. *Macromolecules* **2017**, *50*, 2923–2931. [CrossRef]

26. Luo, F.; Sun, T.L.; Nakajima, T.; Kurokawa, T.; Zhao, Y.; Bin Ihsan, A.; Guo, H.L.; Li, X.F.; Gong, J.P. Crack Blunting and Advancing Behaviors of Tough and Self-healing Polyampholyte Hydrogel. *Macromolecules* **2014**, *47*, 6037–6046. [CrossRef]

27. Nakajima, T. Generalization of the sacrificial bond principle for gel and elastomer toughening. *Polym. J.* **2017**, *49*, 477–485. [CrossRef]

28. Su, E.; Okay, O. Polyampholyte hydrogels formed via electrostatic and hydrophobic interactions. *Eur. Polym. J.* **2017**, *88*, 191–204. [CrossRef]

29. Dyakonova, M.A.; Stavrouli, N.; Popescu, M.T.; Kyriakos, K.; Grillo, I.; Philipp, M.; Jaksch, S.; Tsitsilianis, C.; Papadakis, C.M. Physical Hydrogels via Charge Driven Self-Organization of a Triblock Polyampholyte—Rheological and Structural Investigations. *Macromolecules* **2014**, *47*, 7561–7572. [CrossRef]

30. Chen, Y.Y.; Shull, K.R. High-Toughness Polycation Cross-Linked Triblock Copolymer Hydrogels. *Macromolecules* **2017**, *50*, 3637–3646. [CrossRef]

31. Luo, F.; Sun, T.L.; Nakajima, T.; King, D.R.; Kurokawa, T.; Zhao, Y.; Bin Ihsan, A.; Li, X.F.; Guo, H.L.; Gong, J.P. Strong and Tough Polyion-Complex Hydrogels from Oppositely Charged Polyelectrolytes: A Comparative Study with Polyampholyte Hydrogels. *Macromolecules* **2016**, *49*, 2750–2760. [CrossRef]

32. Bin Ihsan, A.; Sun, T.L.; Kurokawa, T.; Karobi, S.N.; Nakajima, T.; Nonoyama, T.; Roy, C.K.; Luo, F.; Gong, J.P. Self-Healing Behaviors of Tough Polyampholyte Hydrogels. *Macromolecules* **2016**, *49*, 4245–4252. [CrossRef]

33. Na, Y.H. Double network hydrogels with extremely high toughness and their applications. *Korea-Aust. Rheol. J.* **2013**, *25*, 185–196. [CrossRef]

34. Gong, J.P.; Katsuyama, Y.; Kurokawa, T.; Osada, Y. Double-network hydrogels with extremely high mechanical strength. *Adv. Mater.* **2003**, *15*, 1155–1158. [CrossRef]

35. Sun, T.L.; Luo, F.; Kurokawa, T.; Karobi, S.N.; Nakajima, T.; Gong, J.P. Molecular structure of self-healing polyampholyte hydrogels analyzed from tensile behaviors. *Soft Matter* **2015**, *11*, 9355–9366. [CrossRef] [PubMed]

36. Cui, K.P.; Sun, T.L.; Kurokawa, T.; Nakajima, T.; Nonoyama, T.; Chen, L.; Gong, J.P. Stretching-induced ion complexation in physical polyampholyte hydrogels. *Soft Matter* **2016**, *12*, 8833–8840. [CrossRef] [PubMed]

37. Fang, J.; Mehlich, A.; Koga, N.; Huang, J.Q.; Koga, R.; Gao, X.Y.; Hu, C.G.; Jin, C.; Rief, M.; Kast, J.; et al. Forced protein unfolding leads to highly elastic and tough protein hydrogels. *Nat. Commun.* **2013**, *4*, 2974. [CrossRef] [PubMed]

38. Byette, F.; Laventure, A.; Marcotte, I.; Pellerin, C. Metal-Ligand Interactions and Salt Bridges as Sacrificial Bonds in Mussel Byssus-Derived Materials. *Biomacromolecules* **2016**, *17*, 3277–3286. [CrossRef] [PubMed]

39. Huang, W.; Duan, H.D.; Zhu, L.P.; Li, G.Q.; Ban, Q.; Lucia, L.A. A semi-interpenetrating network polyampholyte hydrogel simultaneously demonstrating remarkable toughness and antibacterial properties. *New J. Chem.* **2016**, *40*, 10520–10525. [CrossRef]

40. Dobbins, S.C.; McGrath, D.E.; Bernards, M.T. Nonfouling Hydrogels Formed from Charged Monomer Subunits. *J. Phys. Chem. B* **2012**, *116*, 14346–14352. [CrossRef] [PubMed]

41. Schroeder, M.E.; Zurick, K.M.; McGrath, D.E.; Bernards, M.T. Multifunctional Polyampholyte Hydrogels with Fouling Resistance and Protein Conjugation Capacity. *Biomacromolecules* **2013**, *14*, 3112–3122. [CrossRef] [PubMed]

42. Matsumura, K.; Hayashi, F.; Nagashima, T.; Hyon, S.H. Long-term cryopreservation of human mesenchymal stem cells using carboxylated poly-l-lysine without the addition of proteins or dimethyl sulfoxide. *J. Biomater. Sci.-Polym. Ed.* **2013**, *24*, 1484–1497. [CrossRef] [PubMed]

43. Matsumura, K.; Kawamoto, K.; Takeuchi, M.; Yoshimura, S.; Tanaka, D.; Hyon, S.H. Cryopreservation of a Two-Dimensional Monolayer Using a Slow Vitrification Method with Polyampholyte to Inhibit Ice Crystal Formation. *ACS Biomater. Sci. Eng.* **2016**, *2*, 1023–1029. [CrossRef]

44. Rajan, R.; Jain, M.; Matsumura, K. Cryoprotective properties of completely synthetic polyampholytes via reversible addition-fragmentation chain transfer (RAFT) polymerization and the effects of hydrophobicity. *J. Biomater. Sci.-Polym. Ed.* **2013**, *24*, 1767–1780. [CrossRef] [PubMed]

45. Rajan, R.; Hayashi, F.; Nagashima, T.; Matsumura, K. Toward a Molecular Understanding of the Mechanism of Cryopreservation by Polyampholytes: Cell Membrane Interactions and Hydrophobicity. *Biomacromolecules* **2016**, *17*, 1882–1893. [CrossRef] [PubMed]

46. Jain, M.; Rajan, R.; Hyon, S.H.; Matsumura, K. Hydrogelation of dextran-based polyampholytes with cryoprotective properties via click chemistry. *Biomater. Sci.* **2014**, *2*, 308–317. [CrossRef]

47. Jain, M.; Matsumura, K. Thixotropic injectable hydrogel using a polyampholyte and nanosilicate prepared directly after cryopreservation. *Mater. Sci. Eng. C-Mater. Biol. Appl.* **2016**, *69*, 1273–1281. [CrossRef] [PubMed]

48. Das, E.; Matsumura, K. Tunable Phase-Separation Behavior of Thermoresponsive Polyampholytes Through Molecular Design. *J. Polym. Sci. Part A-Polym. Chem.* **2017**, *55*, 876–884. [CrossRef]

49. Ahmed, S.; Hayashi, F.; Nagashima, T.; Matsumura, K. Protein cytoplasmic delivery using polyampholyte nanoparticles and freeze concentration. *Biomaterials* **2014**, *35*, 6508–6518. [CrossRef] [PubMed]

50. Ahmed, S.; Fujitab, S.; Matsumura, K. Enhanced protein internalization and efficient endosomal escape using polyampholyte-modified liposomes and freeze concentration. *Nanoscale* **2016**, *8*, 15888–15901. [CrossRef] [PubMed]

51. Eckmann, D.M.; Composto, R.J.; Tsourkas, A.; Muzykantov, V.R. Nanogel carrier design for targeted drug delivery. *J. Mater. Chem. B* **2014**, *2*, 8085–8097. [CrossRef] [PubMed]

52. Barcellona, M.N.; Johnson, N.; Bernards, M.T. Characterizing Drug Release from Nonfouling Polyampholyte Hydrogels. *Langmuir* **2015**, *31*, 13402–13409. [CrossRef] [PubMed]

53. Kudaibergenov, S.E.; Tatykhanova, G.S.; Klivenko, A.N. Complexation of macroporous amphoteric cryogels based on *N*,*N*-dimethylaminoethyl methacrylate and methacrylic acid with dyes, surfactant, and protein. *J. Appl. Polym. Sci.* **2016**, *133*. [CrossRef]

54. Rudov, A.A.; Gelissen, A.P.H.; Lotze, G.; Schmid, A.; Eckert, T.; Pich, A.; Richtering, W.; Potemkin, I.I. Intramicrogel Complexation of Oppositely Charged Compartments As a Route to Quasi-Hollow Structures. *Macromolecules* **2017**, *50*, 4435–4445. [CrossRef]

55. Mishra, R.K.; Ramasamy, K.; Ban, N.N.; Majeed, A.B.A. Synthesis of poly 3-(methacryloylamino)propyl trimethylammonium chloride-*co*-methacrylic acid copolymer hydrogels for controlled indomethacin delivery. *J. Appl. Polym. Sci.* **2013**, *128*, 3365–3374. [CrossRef]

56. Cao, Z.F.; Jin, Y.; Miao, Q.; Ma, C.Y.; Zhang, B. Preparation and properties of a dually responsive hydrogels based on polyampholyte for oral delivery of drugs. *Polym. Bull.* **2013**, *70*, 2675–2689. [CrossRef]

57. Sankar, R.M.; Meera, K.M.S.; Samanta, D.; Jithendra, P.; Mandal, A.B.; Jaisankar, S.N. The pH-sensitive polyampholyte nanogels: Inclusion of carbon nanotubes for improved drug loading. *Colloids Surf. B-Biointerfaces* **2013**, *112*, 120–127. [CrossRef] [PubMed]

58. Wang, H.C.; Grolman, J.M.; Rizvi, A.; Hisao, G.S.; Rienstra, C.M.; Zimmerman, S.C. pH-Triggered Release from Polyamide Microcapsules Prepared by Interfacial Polymerization of a Simple Diester Monomer. *ACS Macro Lett.* **2017**, *6*, 321–325. [CrossRef]

59. Schulze, N.; Tiersch, B.; Zenke, I.; Koetz, J. Polyampholyte-tuned lyotrop lamellar liquid crystalline systems. *Colloid Polym. Sci.* **2013**, *291*, 2551–2559. [CrossRef]

60. Ekici, S.; Tetik, A. Development of polyampholyte hydrogels based on laponite for electrically stimulated drug release. *Polym. Int.* **2015**, *64*, 335–343. [CrossRef]

61. Ali, S.A.; Al-Muallem, H.A.; Al-Hamouz, O.; Estaitie, M.K. Synthesis of a novel zwitterionic bisphosphonate cyclopolymer containing residues of alendronic acid. *React. Funct. Polym.* **2015**, *86*, 80–86. [CrossRef]

62. Asayama, S.; Seno, K.; Kawakami, H. Synthesis of Carboxymethyl Poly(1-vinylimidazole) as a Polyampholyte for Biocompatibility. *Chem. Lett.* **2013**, *42*, 358–360. [CrossRef]

63. Ladika, M.; Kalantar, T.H.; Shao, H.; Dean, S.L.; Harris, J.K.; Sheskey, P.J.; Coppens, K.; Balwinski, K.M.; Holbrook, D.L. Polyampholyte Acrylic Latexes for Tablet Coating Applications. *J. Appl. Polym. Sci.* **2014**, *131*. [CrossRef]

*gels*

MDPI

*Review*

# Microfluidic Spun Alginate Hydrogel Microfibers and Their Application in Tissue Engineering

Tao Sun *, Xingfu Li, Qing Shi, Huaping Wang, Qiang Huang and Toshio Fukuda

Beijing Advanced Innovation Center for Intelligent Robots and Systems, Beijing Institute of Technology,
5 South Zhongguancun Street, Haidian District, Beijing 10081, China; ilixingfu@163.com (X.L.);
shiqing@bit.edu.cn (Q.S.); wanghuaping@bit.edu.cn (H.W.); qhuang@bit.edu.cn (Q.H.);
tofukuda@nifty.com (T.F.)
* Correspondence: 3120120061@bit.edu.cn

Received: 9 February 2018; Accepted: 24 March 2018; Published: 23 April 2018

**Abstract:** Tissue engineering is focusing on processing tissue micro-structures for a variety of applications in cell biology and the "bottom-up" construction of artificial tissue. Over the last decade, microfluidic devices have provided novel tools for producing alginate hydrogel microfibers with various morphologies, structures, and compositions for cell cultivation. Moreover, microfluidic spun alginate microfibers are long, thin, and flexible, and these features facilitate higher-order assemblies for fabricating macroscopic cellular structures. In this paper, we present an overview of the microfluidic spinning principle of alginate hydrogel microfibers and their application as micro-scaffolds or scaffolding elements for 3D assembly in tissue engineering.

**Keywords:** microfluidic spinning; alginate hydrogel microfibers; 3D assembly; tissue engineering

## 1. Introduction

Organ transplantation is a well-established therapy for patients suffering from organs failure or damage; however, the long donors-waiting process causes the death of a number of patients, because the source of human organs is extremely limited [1]. To solve such a problem, tissue engineering approaches are trying to create artificial tissues as functional substitutes of human organs. In vivo, multiple cells are spatially arranged with a defined distribution; therefore, replication of such cell arrangement in vitro is critical to achieving the tissue regeneration in vitro [2]. Biomaterials-based scaffolds containing cells take an important extracellular matrix (ECM)-mimicked role to facilitate tissue-like cells arrangement and can further guide cells spreading, proliferation, and differentiation to form artificial tissues with specific shapes. However, traditional scaffolds lack the ability to control the behavior of a single cell and to generate large-scale vascularized tissue [3]. Microfabrication technology is being developed to reduce the size of the scaffolds into micro/nano-scale for tissue regeneration. On the one hand, these micro/nanoscaffolds can be processed with specific topographies and co-culturing cellular patterns for controlling cell behaviors; on the other hand, they can be employed as cell-carriers to achieve precise spatial positioning for the bottom-up construction of the vascularized tissues [4].

A variety of hierarchical scaffolds with controlled morphology and porosity have been engineered by various micro/nanoscaffolds, including polyethylene glycol (PEG) blocks, microdroplets, and micro/nanofibers et al. [5–7]. Among these hierarchical scaffolds, fabricating fibrous structures is one promising component, since micro/nanomicrofibers as scaffolding elements enable the engineered scaffolds to provide physical, chemical, and biological cues to regulate cellular behaviors [8,9]. Electrospinning and microfluidic spinning are two main methods to process micro/nanofibers. Electrospinning produces nanofibers by solidifying a charged polymer jet, and the fabricated nanofibers can be collected together to form net-like structures with nano-scale fiber diameter, high porosity, and stable mechanical properties. The net-like structures have been taken as the promising scaffolds for various tissue applications, such as regeneration of bone, cartilage, muscle, and vessel blood, et al. [10]. Compared with a high DC voltage for spinning nanofibers, microfluidic spinning based on microchannels processed by micro-electromechanical system (MEMs) technology can provide a milder spinning condition, and some unique characters are simultaneously involved in the fabricated microfibers, including cell encapsulation; the spatiotemporal control of microfiber the shape, size, and composition; and manipulation for the single microfiber [11]. Moreover, alginate as the main spinning material has the advantage of rapidly and simply forming gels with relatively stable mechanical properties and biocompatibility [12]. Therefore, microfluidic-spun alginate microfibers are a promising micro-scaffold with great application potential in tissue engineering.

## 2. Microfluidic Spinning Method

Relative to fluidic gravity and inertia, the fluid viscosity and surface tension affect the fluid flow behavior in the microchannel. Such change facilitates the generation of laminar flow in a simple microchannel if the Reynolds number $R_e$ is small [13]. Different solutions form multi-laminar flows in contact with one another, and ions rapidly exchange between different laminar flow. In general, microfluidic spinning methods are dependent on the exchange of $Ca^{2+}$ ions between alginate flow and $CaCl_2$ flow [14]. Alginate microfibers with different morphologies and functions can be spun by three kinds of flow dynamic systems according to the different structures of the microchannels.

### 2.1. Parallel Laminar Flows

Parallel laminar flows are generated in rectangular polydimethylsiloxane (PDMS) microchannel with a uniform thickness. The microchannel is fabricated by standard soft lithography and replica molding techniques, and its height is decided by the coating thickness of SU-8 [15]. Typical microfluidic device consists of three inlets and a long gelation microchannel, as shown in Figure 1a. Alginate solution is injected from the middle inlet by syringe pump, and then is squeezed by $CaCl_2$ solution injected from two side inlets. Afterwards, sandwiched laminar flows are formed in the gelation microchannel. In laminar flows, $Ca^{2+}$ ions in $CaCl_2$ flow rapidly diffuse in the horizontal direction of the microchannel, and the alginate flow is gradually transformed into alginate microfibers. The cross-sectional shape of microfibers is a rounded rectangle, and the cross-sectional area can be modified by changing the flow rate of all solution [16]. Moreover, a thickener, such as dextran, can be added in $CaCl_2$ solution to balance the viscosities of the solutions, which promotes the modification of the cross-sectional area. In addition, a buffer flow can be injected to form buffer flow between alginate and $CaCl_2$ flows [17]. On the one hand, buffer flow moderates rapid cross-linking reaction to prevent the formation of gel irregular from blocking the microchannel; on the other hand, accelerating the buffer flow can terminate the spinning process to effectively pinch off the spun alginate microfibers from the gelation microchannel [18].

Alginate solution with different components or various concentrations can be simultaneously introduced to form a parallel multi-alginate flows in the gelation microchannel [19]. However, excessive inlets may waste the device space and expensive syringe pumps. Therefore, the multi-layered distribution channel networks connected to various inlets are employed to divide the different injected solutions into multi-flows and subsequently recombine flows with controlled array pattern [20]. Because of wide cross-section area of the recombined flows, the sufficient cross-linking reaction for solidifying the flows is difficult to achieve when dependent on $CaCl_2$ flow. Therefore, the recombined flows are injected in a dish containing $CaCl_2$ solution. In this case, the cross-sectional area of the spun microfibers is mainly decided by the size of the outlet of gelation microchannel rather than flow rate of the solutions.

## 2.2. Coaxial Laminar Flows

The formation of coaxial laminar flows requires a microfluidic device with coaxial geometry [21]. The coaxial microfluidic device typically consists of an inner tapered glass capillary and an outer cylindrical or square tube, as shown in Figure 1b. Precise alignment and fixing technology is key to coaxially nesting the tapered part of inner capillary in the outer tube. Alginate solution is injected from inner capillary and then is sheathed by $CaCl_2$ solution introduced from outer tube to form coaxial laminar flows. Hydrogel microfibers with circular or flat cross-sectional shapes can be spun responding to the cylindrical and square structure of the inlet, respectively [22,23]. Relative to the horizontal diffusion of $Ca^{2+}$ ions in parallel laminar flows, the gelling direction in coaxial laminar flows is uniform in the cross-section of alginate flows. Furthermore, multi layered coaxial laminar flows can be formed by adding the number of nested capillary [24]. Recently, a multi-coaxial microfluidic device has been widely constructed to generate a core-shell hydrogel microfiber, as shown in Figure 1c [25]. Different core parts of microfibers, including laminar flows of hyaluronic acid, biological cells, ECM protein, linear array of silk fibroin, and oil microdroplets can be sheathed by alginate hydrogel layer as outer protective shell [26–30]. Although such core-shell alginate microfibers can be formed in parallel laiminar flow, the encapsulation of the core parts is unstable [31]. In addition, coaxial flow enables the spun alginate microfiber to be spiraled by an unbalanced fluidic friction induced by its surrounding flow to form helical microfibers [32,33].

Besides glass capillary, PDMS microfluidic device can also be employed to generate a coaxial flow system. A step-like microchannel is characteristic of such PDMS device, which can be constructed by fabricating two consecutive SU-8 molds with different heights and replica molding. By aligning and bonding two step-like microchannel, a coaxial microchannel with different cross-section area can be fabricated. Alginate solution injected from the smaller microchannel can eject into $CaCl_2$ flow injected from the microchannel with the bigger microchannel to form coaxial flow, and the flat alginate microfiber can be spun. In additions, the cylindrical microfibers can also be generated by employing the deflection of free-standing thin PDMS membranes to fabricate a cylindrical-flow PDMS channel [34]. Benefiting from the fabrication ease and stable physical properties of PDMS, sophisticated microchannel architecture can be constructed to involve some unique configuration characteristic into alginate microfibers. By engraving the grooved pattern on the inner surface of inlet, flat alginate fibers with grooves are spun under the sheath effect of $CaCl_2$ flow [35]. Furthermore, a multiply stacked PDMS structure enables the formation of laminar flows in Z direction rather than in XY planar in traditional microchannel, and alginate microfibers with complex cross-sectional shapes can be formed by combining symmetric and coaxial laminar flows simultaneously formed in such microchannel [36].

## 2.3. Valve-Involved Spinning Method

Despite the fact that the above-mentioned laminar flow system enables mass production of alginate microfiber with various components, the distribution of these components in microfibers is decided by the injection position of alginate precursors in advance. For achieving the tunable distribution of components, a valve control system is designed to facilitate the programmable flow control in the spinning process [37]. Specifically, a parabolic PDMS membrane naturally generated in an air hole was installed on the inlet microchannel. Continuous air pressure or vacuum generated in air hole enables the membrane to be deflected or retrieved for 'on-off' switching of flow injection. By installing the valve on each inlet channel and independently controlling it, alginate microfiber with coded compositions can be spun, as shown in Figure 1d. Furthermore, the valve can be operated to control the core of droplets to be distributed either periodically or uniformly throughout the core-shell alginate microfiber [38].

In conclusion, fabricating alginate microfibers with complex compositions and morphology is a hotspot in microfluidic spun method. The complexity of the spun microfiber matches with the microchannel architecture. Rectangular PDMS microchannel is easily fabricated, in which parallel laminar flows can be formed to spun flat microfibers. Although the fabrication of glass microcapillaries requires skill, the generated coaxial flow can create novel core-shell cylindrical and helical microfibers. Furthermore, the most labor-intensive multi-layers PDMS microchannel with specific architectures allows for construct of alginate microfibers with tunable multi-compartments and morphologies.

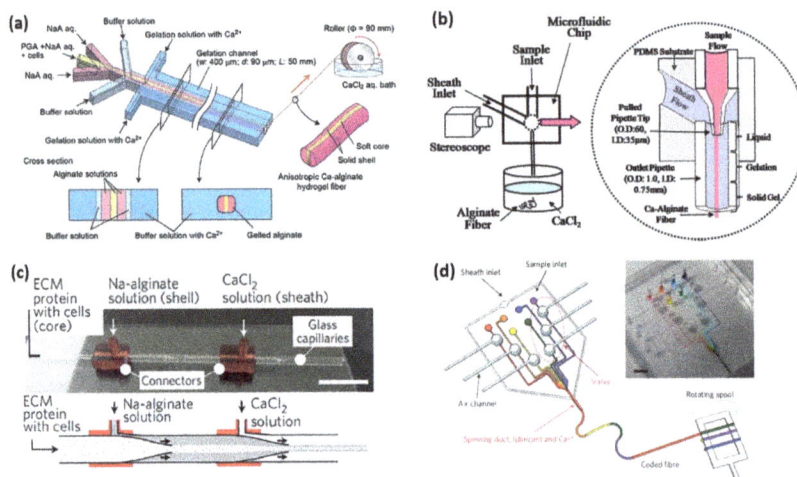

**Figure 1.** Microfluidic spinning system. (**a**) Parallel laminar flows [16]. Adapted with permission from [16]. Copyright 2012 Royal Society of Chemistry. (**b**) Coaxial laminar flows [21]. Adapted with permission from [21]. Copyright 2007 American Chemical Society. (**c**) Multi-coaxial microfluidic device [27]. Adapted with permission from [27]. Copyright 2013 Springer Nature. (**d**) Valve-involved spinning method [38]. Adapted with permission from [38]. Copyright 2011 Springer Nature.

## 3. Alginate Hydrogel Microfibers as Scaffolds for Cell Culture

Laminar flows in microchannels can provide a mild aqueous environment to encapsulate cells into alginate microfibers with high viability, and the encapsulated cells can keep their activity for a long time, since alginate hydrogel provides a similar structure with the extracellular matrices in tissue. The easy encapsulation of cells and inherent biocompatibility enable microfluidic spun alginate microfibers to be widely applied as microscaffolds in tissue engineering.

### 3.1. Three-Dimensional Cell Culture

Since $Ca^{2+}$ ions can generate a high degree of coordination with guluronate blocks of alginate chains, neighboring guluronate blocks of different alginate chains can then be combined to form "egg-box" junctions [12]. Plenty of junctions resemble a three-dimensionally networks to stably fix cells in the cross-linked alginate microfibers. Intrinsically higher porosity and larger pore sizes facilitate the exchange of nutrition and cell-secreted molecules and waste. However, due to the lack of adhesive sites on mammalian cell, pure alginate microfibers are usually employed as a simple mechanical support shell with function of cell immobilization and immunoprotection in vivo [21]. Utilizing the co-coaxial flows, Onoe et al. and Jun et al. and encapsulated the mixture of ECM proteins and pancreatic islet cells as cores into alginate microfibers as support shells [25,39]. ECM proteins can provide a suitable microenvironment to promote the islet cell to behave as they do in vivo. The alginate shell not only allows insulin secreted by islets cells to diffuse out of the fibers into the surrounding tissue, but also to pass and protect cells from immunological attack. In addition, differentiated fat cells encapsulated into such ECM proteins' cores can successfully differentiate into smooth muscle-like cells, inducing the coiling of alginate microfibers, as shown in Figure 2a [40]. The circumferential orientation during cultivation is important for revealing the in vivo-like cell phenotype [41].

For promoting anchorage-dependent cells to interact with alginate hydrogel, other biopolymers have been incorporated with alginate to form composite hydrogel microfibers [12]. By the direct mixing of water-soluble chitosan solution and alginate solution, Lee et al. synthetized microfluidic-spun chitosan-alginate microfibers based on $Ca^{2+}$ ions-induced cross-linked reaction, and the encapsulated human hepatocellular carcinoma (HepG2) cells were more viable than the cells encapsulated in pure alginate hydrogel [42]. Furthermore, the grafting of Arginylglycylaspartic acid (RGD) facilitates alginate hydrogel with higher biocompatibility for cell functional expression. Utilizing the combination of ionic and photoinitiated cross-linking, a RGD-incorporated double-layer of hollow microfibers containing human umbilical vascular endothelial cells (HUVECs) and human osteoblast-like cells (MG63) has been fabricated to form a biomimetic osteon-like structure. The encapsulated cells present an elongated cell morphology indicating excellent interactions between cells and surrounding matrix, which is difficult to achieve in chitosan-alginate microfiber. However, the introduction of $Ca^{2+}$ ions and ionic cross-linking may reduce the mechanical properties of hydrogel microfibers [43]. As an alternative, alginate hydrogel has been regulated by incorporation of methacrylated gelatin (GelMA) containing RGD. Zuo et al. reported use glass microcapillary to transfer such composite materials into hydrogel microfibers with combination of high mechanical moduli and high biocompatibility [44].

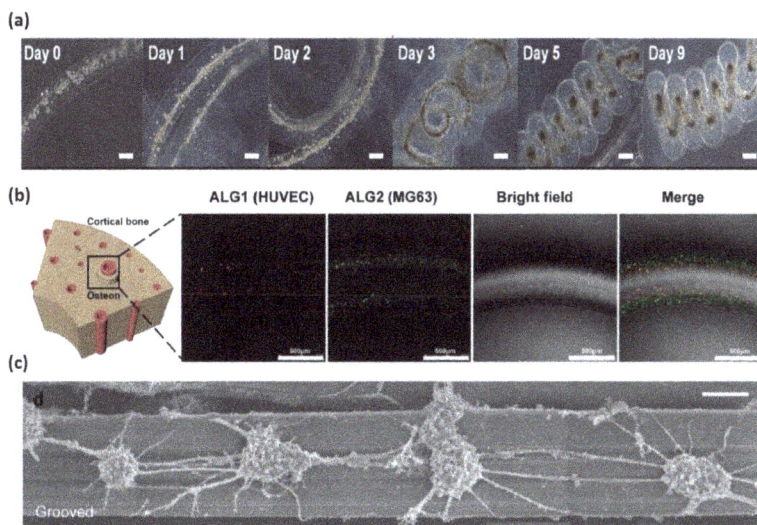

**Figure 2.** Cell-laden alginate hydrogel microfibers. (**a**) Self-assembly of de-differentiated fat cells-contained alginate microfiber [40], adapted with permission from [40]. Copyright 2015 Hsiao et al. Scale bar indicates 100 μm. (**b**) Osteon-like microfibers [43]. (**c**) Alginate microfiber with grooves for aligning nerve cells [35], adapted with permission from [35]. Copyright 2012 John Wiley and Son. Scale bar indicates 50 μm.

## 3.2. Biomimetic Microorganoids

Because of the locating capacity of target fluids and cells on designed sections, multi-laminar flow enables alginate microfibers to be constructed into micro-organoids containing multiple cells with hierarchical structures mimicking nature microtissues, such as hepatic lobules, microvascular micro-scale blood vessel, and osteon. Because hepatocytes may lose their phenotype without the support of the feeder cells, in vitro patterned co-cultured microenvironment is important to maintain hepatocyte functions [45]. Utilizing three laminar flows generated in flat PDMS microchannel, Yamada et al. fabricated sandwiched Ba-alginate hydrogel microfibers, in which hepatocytes at the center are closely encased by Swiss 3T3 cells from two sides [19]. Compared with hepatocytes singly cultured in alginate hydrogel microfibers, hepatic functions including albumin secretion and urea synthesis are significantly enhanced due to such co-cultivation under high oxygen tension. Alginate lyase can enzymatically digest the surrounding hydrogel of the hepatic structure to obtain a micro-organoids mimicking in vivo hepatic cord structures without cells damage. Furthermore, they fabricated a stripe-patterned heterogeneous alginate microfiber with multi-sandwiched structure, which facilitates the relatively large-scale formation of rod-like heterotypic organoids for high retring of liver-specific functions [20].

Cylindrical microchannel can generate multi-layered coaxial laminar flows with circular cross-section. When the laminar flow at center is not involved in gelation reaction with the surrounding alginate flow containing cells, hollow alginate microfibers can be formed to mimic tubular microtissue in vivo [23]. By assembling such hollow alginate microfibers, Gao et al. fabricated large-scale 3D hydrogel structures with built-in microchannels, which enables the encapsulated L929 mouse fibroblasts to show higher cell viability relative to the hydrogel structures without built-in microchannels [46]. To further promote the bioactivity, Jia et al. employed a blend flow including gelatin methacryloyl (GelMA), alginate, and 4-arm poly (ethylene glycol)-tetra-acrylate (PEGTA) to form hollow alginate microfibers by successive ionic and photo crosslinking, and the resulting microfibers allow the spreading and proliferation of the encapsulated vascular cells, which is critical for creating functional vascular-like structure [47]. Furthermore, Xu et al. injected the rat blood into such hollow microfibers with helical structure to achieve their perfusabililty [48]. In additions, Wei et al. engineered hollow microfibers with double RGD-introduced alginate hydrogel layers. Osteocytes and human umbilical cord vein endothelial cells (HUVECs) were respectively encapsulated into inner and outer layer to mimic the structure of osteons, and such osteon-like microfibers enable cells to achieve the enhanced osteogenic and vasculogenic expression, as shown in Figure 2b [43].

### 3.3. Cell Guidance

For constructing linear tissue such as nerve bundles and muscle fibers, cells should be guided to achieve linear alignment during cultivation. The narrow and long morphology enables microfibers as a promising linear scaffold for regeneration of linear tissues. Yamada et al. fabricated solid-soft-solid flat microfibers by forming parallel alginate flows with and without propylene glycol alginate (PGA) [16]. Because of the physical restriction of two side solid region and the surrounding PLL layer for cell expanding, the neuron-like PC12 cells encapsulated in central soft region can be guided to elongate and generate cellular network along the microfiber direction. Furthermore, they designed a vertical micronozzle array structure to spin cylindrical alginate microfibers with eight soft regions uniformly located in the microfiber periphery [49]. Such complicated structure results in a higher ratio of linear colony formation of PC12 cells than the ratio in the proposed sandwiched structure, and the culture of PC12 cells in multi-soft regions facilitates the formation of one-millimeter-long intercellular networks that mimic nerve bundles in vivo.

Nerve cells can be alternatively cultured on the surface of alginate microfibers with high cell density, effective cell connection, and a good observation. Kang et al. fabricated grooved flat alginate microfibers to guide the neurites of neuron cells to form networks on the microfiber surface, as shown in Figure 2c [35]. The morphology of the connected neurite bundles of different neuron cells can be clearly observed. In contrast, most neuron cells aggregated on the edges of smooth microfibers, and the connection between different cells is difficult to be distinguished. The narrow groove bridges as a mechanical anchor are critical for guided cell expanding, and the same effect can be achieved by narrow microfibers with the nearly same size of the groove bridge. We used flat alginate microfibers (horizontal/vertical width < 40 μm) as scaffolding elements to guide fibroblast cells to form toroidal cellular micro-rings for the fabrication of microvascualr-like structure [50]. Compared with cells residing in microfibers, cells adhered on surface of alginate microfibers can be easily guided to form cellular layer with high cell density and uniform distribution.

## 4. Alginate Hydrogel Microfibers as Scaffolding Elements for Higher-Order Tissue

Long and flexible structure and stable mechanical properties facilitate the excellent processability of alginate microfibers. The ease of mixing with other biomaterials promotes the high biocompatibility of the microfibers for cells growth. Microfluidic spinning method further involves turning biomimetic functions into microfibers. These features enable alginate microfibers as promising building blocks for the assembly of 3D macroscopic tissue-like structures. At present, the assembly strategy mainly consists of microfluidic printing and manipulation-based assembly.

### 4.1. Microfluidic Printing

In view of the capacity of the precise arrangement for cellular aggregates and cell-laden hydrogel microblocks, bioprinting becomes an efficient approach to directly construct the 3D cellular constructs with clinically relevant size, arbitrary shape, and vascularized architecture [51]. Compared with the pressure-induced generation of beads or lines of materials containing cells, alginate microfibers as printing ink can be generated in a relatively mild processing microenvironment. Ghorbanian, et al. employed a motorized stage to control the movement of microfluidic device for arranging the spun alginate microfiber [52]. When the movement speed is synchronized with the speed of fiber fabrication, cell-laden alginate microfibers can be smoothly deposited on the surface in the x-direction, and then in y-direction, and so on to form a multilayer net-like structure with high cell viability. Moreover, unique micro-structures generated by microfluidic spinning can be simultaneously involved into the generated net-like structure. Gao et al. offer a 3D bioprinting strategy based on hollow alginate microfiber spun by a coaxial nozzle as printing head [46]. With a Z-shape platform, the layer-by-layer fabrication can be achieved to form a cell-laden hydrogel structure with built-in microchannels, and cell culture can be perfused through the built-in microchannels showing the feasibility of nutrients flow, as shown in Figure 3a. Jia et al. improved the gelation materials form pure alginate into cell-responsive materials consisting of GelMA, alginate, and PEGTA [47]. The encapsulated endothelial and stem cells can spread and proliferate in the wall of the arranged hollow microfibers to form biomimetic built-in microchannels, leading to the formation of biological vasculature, as shown in Figure 3b. However, due to relatively higher mechanical properties, alginate microfibers are difficult to be arbitrarily arranged as the popular printing ink. Therefore, except for net-like structures, it is still a challenge to print other complicated structure mimicking natural tissue.

### 4.2. Guided-Assembly Method

The feasibility to handle single microfiber provides an alternative way to implement the higher-order assembly of alginate microfibers. With the guidance of weaving machine, weaving-based assembly of textile fibers can be widely applied in clothing industry. Long and flexible structure features enable alginate microfibers to have a potential to be weaved; however, new challenges to achieve such weaving assembly are generated due to the manipulation difficulty induced by micro-scale size, cell encapsulation, and weak mechanical properties relative to textile fibers. Onoe et al. developed a capillary-based "microfluidic handling" method to manipulate alginate microfibers [27]. The alginate microfiber sucked into capillary can be held by fluid resistance without any mechanical attachment. The clamped microfiber can move with the moving capillary and can be tensioned to form a bridge between two capillary holding two microfiber ends, respectively. Utilizing such bridge-shaped architecture, microfibers can be weaved to create a 3D net-like structure, and the net can keep stable by the friction among microfibers without secondary cross-linking reaction applied in microfluidic "printing" method, as shown in Figure 3c. The excess concentration of $Ca^{2+}$ induced by secondary cross-linking reactions may induced cell death. When the medium is pushed out, the clamped end of the microfibers can be released from the capillary, and the fabric-like cellular structure can be obtained to mimic in vivo neuronal pathways in the brain.

Magnetic nanoparticles can be uniformly dispersed into alginate solution by ultrasonic vibration, and then be encapsulated into alginate microfibers by microfluidic spinning. The egg-box model of cross-linking enables the Ca-alginate gel to generate strong adhesive forces to capture the moving MNPs; therefore, magnetic alginate microfiber can be moved towards the external magnetic field. Based on such enhanced controllability, a series of magnetic guidance and no-contact manipulation strategies are developed for potential application in 3D cellular assembly. He et al. employed magnetic arrays to construct microfiber 2D letters and 3D textile-like patterns without the requirement of specific peripheral equipments [30]. The superparamagnetic properties of magnetic nanoparticles enable alginate microfibers to be stably adhered on the surface of the arrays, which is critical to keeping the patterned structure in the solution environments. We further miniatured the magnetic arrays by using an array of iron wires connected on the magnet surface. The iron wire can focus the magnetic field into the top of array, which facilitates more precise guidance for microfiber pattern relative to the magnet arrays [53]. Furthermore, magnetic response of alginate microfibers facilitates fabrication of more complicated structure using microfluidic "printing" method. We put a ring magnet under a support model (semi-sphere or curved semi-cylinder model) adhered on the bottom surface of dish filled with PBS. Because of the directional attraction of magnet, microfibers can neglect the effect of buoyancy to directly deposit on the surface of the models. Layer-by-layer microfibers can be assembled along the outline of the model with the movement of microfluidic device. After secondary cross-linking and removing the magnet, 3D cell-laden structure with the shapes of dome and curved arc can be obtained to mimic in vivo diaphragm beneath the lungs and curved vascular structure.

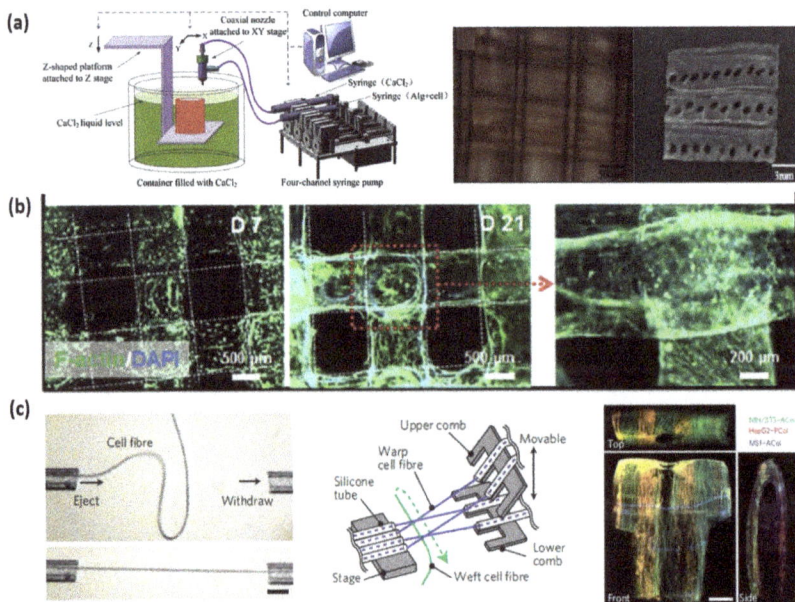

**Figure 3.** Higher-order assemblies for tissue engineering. (**a**) A printed cuboid structure consisting of 6 layers of hollow microfibers [46]. Adapted with permission from [46]. Copyright 2015 Elsevier Ltd. (**b**) Bioprinted hollow microfibers encapsulating vascular cells [47]. Adapted with permission from [47]. Copyright 2016 Elsevier Ltd. (**c**) Centimetre-scale woven macroscopic tissue [27], adapted with permission from [27]. Copyright 2013 Spring Nature. Scale bars represent 1 mm.

## 5. Conclusions and Future Perspective

Microfluidic spinning method facilitates the production of alginate hydrogel microfibers with different morphologies and compositions to promote cell proliferation. Different cells, including fibroblast, endothelial cells, smooth muscle cells, nerve cells, hepatic cells, et al., have successfully encapsulated in or covered the alginate microfibers as microscaffolds for the research of cell biology and tissue engineering. However, due to the poor controllability of alginate microfibers, higher-order assembly is still challenging for the construction of macroscope cellular structure mimicking in vivo tissue. Therefore, the manipulation methods and assembly strategies should be further studied for construction of complex cellular structures.

**Acknowledgments:** This work is supported by the China Postdoctoral Science Foundation (2016M600932) and the National Natural Science Foundation of China (61703045 and 61520106011).

**Conflicts of Interest:** The authors declare no conflict of interest.

## References

1. Langer, R.; Vacanti, J.P. Tissue engineering. *Science* **1993**, *260*, 920–926. [CrossRef] [PubMed]
2. Mironov, V.; Visconti, R.P.; Kasyanov, V.; Forgacs, G.; Drake, C.J.; Markwald, R.R. Organ printing: Tissue spheroids as building blocks. *Biomaterials* **2009**, *20*, 2164–2174. [CrossRef] [PubMed]
3. Vacanti, C.A. The history of tissue engineering. *J. Cell. Mol. Med.* **2006**, *10*, 569–576. [CrossRef] [PubMed]
4. Nichol, J.W.; Khademhosseini, A. Modular tissue engineering: Engineering biological tissues from the bottom up. *Soft Matter* **2009**, *5*, 1312–1319. [CrossRef] [PubMed]
5. Xu, F.; Wu, C.A.M.; Rengarajan, V.; Finley, T.D.; Keles, H.O.; Sung, Y.R.; Li, B.Q.; Gurkan, U.A.; Demirci, U. Three-dimensional magnetic assembly of microscale hydrogels. *Adv. Mater.* **2011**, *23*, 4254–4260. [CrossRef] [PubMed]
6. Park, D.Y.; Mun, C.H.; Kang, E.; No, D.Y.; Ju, J.; Lee, S.H. One-stop microfiber spinning and fabrication of a fibrous cell-encapsulated scaffold on a single microfluidic platform. *Biofabrication* **2014**, *6*, 024108. [CrossRef] [PubMed]
7. Matsunaga, Y.T.; Morimoto, Y.; Takeuchi, S. Molding cell beads for rapid construction of macroscopic 3D tissue architecture. *Adv. Mater.* **2011**, *12*, H90–H94. [CrossRef] [PubMed]
8. Nain, A.S.; Philippi, J.A.; Sitti, M.; MacKrell, J.; Campbell, P.G.; Amon, C. Control of cell behavior by aligned micro/nanofibrous biomaterial scaffolds fabricated by spinneret-based tunable engineered parameters (STEP) technique. *Small* **2008**, *4*, 1153–1159. [CrossRef] [PubMed]
9. Berry, S.M.; Warren, S.P.; Hilgart, D.A.; Schworer, A.T.; Pabba, S.; Gobin, A.S.; Cohn, R.W.; Keynton, R.S. Endothelial cell scaffolds generated by 3D direct writing of biodegradable polymer microfibers. *Biomaterials* **2011**, *32*, 1872–1879. [CrossRef] [PubMed]
10. Cheng, J.; Jun, Y.; Qin, J.H.; Lee, S.H. Electrospinning versus microfluidic spinning of functional fibers for biomedical applications. *Biomaterials* **2017**, *114*, 121–143. [CrossRef] [PubMed]
11. McNamara, M.C.; Sharifi, F.; Wrede, A.H.; Kimlinger, D.F.; Thomas, D.G.; Vander Wiel, J.B.; Vander Wiel, J.B.; Chen, Y.F.; Montazami, R.; Hashemi, N.N. Microfibers as physiologically relevant platforms for creation of 3D cell cultures. *Macromol. Biosci.* **2017**, *17*, 1700279. [CrossRef] [PubMed]
12. Lee, K.Y.; Mooney, D.J. Alginate: Properties and biomedical applications. *Prog. Polym. Sci.* **2012**, *37*, 106–126. [CrossRef] [PubMed]
13. Atencia, J.; Beebe, D.J. Controlled microfluidic interfaces. *Nature* **2005**, *437*, 648–655. [CrossRef] [PubMed]
14. Knight, J.B.; Vishwanath, A.; Brody, J.P.; Austin, R.H. Hydrodynamic focusing on a silicon chip: Mixing nanoliters in microseconds. *Phys. Rev. Lett.* **1998**, *80*, 3863–3866. [CrossRef]
15. Ng, J.M.K.; Gitlin, I.; Stroock, A.D.; Whitesides, G.M. Components for integrated poly (dimethylsiloxane) microfluidic systems. *Electrophoresis* **2002**, *23*, 3461–3473. [CrossRef]
16. Yamada, M.; Sugaya, S.; Naganuma, Y.; Seki, M. Microfluidic synthesis of chemically and physically anisotropic hydrogel microfibers for guided cell growth and networking. *Soft Matter* **2012**, *8*, 3122–3130. [CrossRef]

17. Lin, Y.S.; Huang, K.S.; Yang, C.H.; Wang, C.Y.; Yang, Y.S.; Hsu, H.C.; Liao, Y.J.; Tsai, C.W. Microfluidic synthesis of microfibers for magnetic-responsive controlled drug release and cell culture. *PLoS ONE* **2012**, *7*, e33184. [CrossRef] [PubMed]

18. Sun, T.; Huang, Q.; Shi, Q.; Wang, H.P.; Liu, X.M.; Seki, M.; Nakajima, M.; Fukuda, T. Magnetic assembly of microfluidic spun alginate microfibers for fabricating three-dimensional cell-laden hydrogel constructs. *Microfluid. Nanofluid.* **2015**, *19*, 1169–1180. [CrossRef]

19. Yamada, M.; Utoh, R.; Ohashi, K.; Tatsumi, K.; Yamato, M.; Okano, T.; Seki, M. Controlled formation of heterotypic hepatic micro-organoids in anisotropic hydrogel microfibers for long-term preservation of liver-specific functions. *Biomaterials* **2012**, *33*, 304–8315. [CrossRef] [PubMed]

20. Kobayashi, A.; Yamakoshi, K.; Yajima, Y.; Utoh, R.; Yamada, M.; Seki, M. Preparation of stripe-patterned heterogeneous hydrogel sheets using microfluidic devices for high-density coculture of hepatocytes and fibroblasts. *J. Biosci. Bioeng.* **2013**, *116*, 761–767. [CrossRef] [PubMed]

21. Shin, S.; Park, J.Y.; Lee, J.Y.; Park, H.; Park, Y.D.; Lee, K.B.; Whang, C.M.; Lee, S.H. "On the fly" continuous generation of alginate fibers using a microfluidic device. *Langmuir* **2007**, *23*, 9104–9108. [CrossRef] [PubMed]

22. Yu, Y.; Wei, W.B.; Wang, Y.Q.; Xu, C.; Guo, Y.Q.; Qin, J.H. Simple Spinning of Heterogeneous Hollow Microfibers on Chip. *Adv. Mater.* **2016**, *23*, 6649–6655. [CrossRef] [PubMed]

23. Lee, K.H.; Shin, S.J.; Park, Y.; Lee, S.H. Synthesis of cell-laden alginate hollow fibers using microfluidic chips and microvascularized tissue-engineering applications. *Small* **2009**, *5*, 1264–1268. [CrossRef] [PubMed]

24. Cheng, Y.; Zheng, F.Y.; Lu, J.; Shang, L.R.; Xie, Z.Y.; Zhao, Y.J.; Chen, Y.P.; Gu, Z.Z. Bioinspired multicompartmental Microfibers from Microfluidics. *Adv. Mater.* **2014**, *26*, 5184–5190. [CrossRef] [PubMed]

25. Chaurasia, A.S.; Sajjadi, S. Flexible asymmetric encapsulation for dehydration-responsive hybrid microfibers. *Small* **2016**, *12*, 4146–4155. [CrossRef] [PubMed]

26. Meng, Z.J.; Wang, W.; Xie, R.; Ju, X.J.; Liu, Z.; Chu, L.Y. Microfluidic generation of hollow ca-alginate microfibers. *Lab Chip* **2016**, *16*, 2673–2681. [CrossRef] [PubMed]

27. Onoe, H.; Okitsu, T.; Itou, A.; Kato-Negishi, M.; Gojo, R.; Kiriya, D.; Sato, K.; Miura, S.; Iwanaga, S.; Kuribayashi-Shigetomi, K.; et al. Metre-long cell-laden microfibres exhibit tissue morphologies and functions. *Nat. Mater.* **2013**, *12*, 584–590. [CrossRef] [PubMed]

28. Zhu, Y.J.; Wang, L.; Yin, F.C.; Yu, Y.; Wang, Y.Q.; Liu, H.; Wang, H.; Sun, N.; Liu, H.T.; Qin, J.H. A hollow fiber system for simple generation of human brain organoids. *Intergr. Biol. (Camb.)* **2017**, *9*, 774–781. [CrossRef] [PubMed]

29. Cheng, J.; Park, D.; Jun, Y.; Lee, J.; Hyun, J.; Lee, S.H. Biomimetic spinning of silk fibers and in situ cell encapsulation. *Lab Chip* **2016**, *16*, 2654–2661. [CrossRef] [PubMed]

30. He, X.H.; Wang, W.; Liu, Y.M.; Jiang, M.Y.; Wu, F.; Deng, K.; Liu, Z.; Ju, X.J.; Xie, R.; Chu, L.Y. Microfluidic Fabrication of Bio-Inspired Microfibers with Controllable Magnetic Spindle-Knots for 3D Assembly and Water Collection. *ACS Appl. Mater. Interfaces* **2015**, *7*, 17471–17481. [CrossRef] [PubMed]

31. Sun, T.; Hu, C.Z.; Nakajima, M.; Takeuchi, M.; Seki, M.; Yue, T.; Shi, Q.; Fukuda, T. On-chip fabrication and magnetic force estimation of peapod-like hybrid microfibers using a microfluidic device. *Microfluid. Nanofluid.* **2015**, *18*, 1177–1187. [CrossRef]

32. Yu, Y.R.; Fu, F.F.; Shang, L.R.; Cheng, Y.; Gu, Z.Z.; Zhao, Y.J. Bioinspired Helical Microfibers from Microfluidics. *Adv. Mater.* **2017**, *29*, 1605765. [CrossRef] [PubMed]

33. Totton, S.; Takeuchi, S. Formation of liquid rope coils in a coaxial microfluidic device. *RSC Adv.* **2015**, *5*, 33691–33695.

34. Kang, E.; Shin, S.J.; Lee, K.H.; Lee, S.H. Novel PDMS cylindrical channels that generate coaxial flow, and application to fabrication of microfibers and particles. *Lab Chip* **2010**, *10*, 1856–1861. [CrossRef] [PubMed]

35. Kang, E.; Choi, Y.Y.; Chae, S.K.; Moon, J.H.; Chang, J.Y.; Lee, S.H. Microfluidic Spinning of Flat Alginate Fibers with Grooves for Cell-Aligning Scaffolds. *Adv. Mater.* **2012**, *24*, 4271–4277. [CrossRef] [PubMed]

36. Yoon, D.H.; Kobayashi, K.; Tanaka, D.; Sekiguchi, T.; Shoji, S. Simple microfluidic formation of highly heterogeneous microfibers using a combination of sheath units. *Lab Chip* **2017**, *17*, 1481–1486. [CrossRef] [PubMed]

37. Yu, Y.; Wen, H.; Ma, J.Y.; Lykkemark, S.; Xu, H.; Qin, J.H. Flexible Fabrication of Biomimetic Bamboo-Like Hybrid Microfibers. *Adv. Mater.* **2014**, *26*, 2494–2499. [CrossRef] [PubMed]

38. Kang, E.; Jeong, G.S.; Choi, Y.Y.; Lee, K.H.; Khademhosseini, A.; Lee, S.H. Digitally tunable physicochemical coding of material composition and topography in continuous microfibres. *Nat. Mater.* **2011**, *10*, 877–883. [CrossRef] [PubMed]

39. Jun, Y.; Kim, M.J.; Hwang, Y.H.; Jeon, E.A.; Kang, A.R.; Lee, S.H.; Lee, D.Y. Microfluidics-generated pancreatic islet microfibers for enhanced immunoprotection. *Biomaterials* **2013**, *34*, 8122–8130. [CrossRef] [PubMed]

40. Hsiao, A.Y.; Okitsu, T.; Onoe, H.; Kiyosawa, M.; Teramae, H.; Iwanaga, S.; Kazama, T.; Matsumoto, T.; Takeuchi, S. Smooth Muscle-Like Tissue Constructs with Circumferentially Oriented Cells Formed by the Cell Fiber Technology. *PLoS ONE* **2015**, *10*, e0119010. [CrossRef] [PubMed]

41. Choi, J.S.; Piao, Y.; Seo, T.S. Circumferential alignment of vascular smooth muscle cells in a circular microfluidic channel. *Biomaterials* **2014**, *35*, 63–70. [CrossRef] [PubMed]

42. Lee, B.R.; Lee, K.H.; Kang, E.; Kim, D.S.; Lee, S.H. Microfluidic wet spinning of chitosan-alginate microfibers and encapsulation of HepG2 cells in fibers. *Biomicrofluidics* **2011**, *5*, 022208. [CrossRef] [PubMed]

43. Wei, D.; Sun, J.; Bolderson, J.; Zhong, M.L.; Dalby, M.J.; Cusack, M.; Yin, H.B.; Fan, H.S.; Zhang, X.D. Continuous Fabrication and Assembly of Spatial Cell-Laden Fibers for a Tissue-Like Construct via a Photolithographic-Based Microfluidic Chip. *ACS Appl. Mater. Interfaces* **2017**, *9*, 14606–14617. [CrossRef] [PubMed]

44. Zuo, Y.C.; He, X.H.; Yang, Y.; Wei, D.; Sun, J.; Zhong, M.L.; Xie, R.; Fan, H.S.; Zhang, X.D. Microfluidic-based generation of functional microfibers for biomimetic complex tissue construction. *Acta Biomater.* **2016**, *38*, 153–162. [CrossRef] [PubMed]

45. Dunn, J.C.; Tompkins, R.G.; Yarmush, M.L. Hepatocytes in collagen sandwich: Evidence for transcriptional and translational regulation. *J. Cell Biol.* **1992**, *116*, 1043–1053. [CrossRef] [PubMed]

46. Gao, Q.; He, Y.; Fu, J.Z.; Liu, A.; Ma, L. Coaxial nozzle-assisted 3D bioprinting with built-in microchannels for nutrients delivery. *Biomaterials* **2015**, *61*, 203–215. [CrossRef] [PubMed]

47. Jia, W.T.; Gungor-Qzkerim, P.S.; Zhang, Y.S.; Yue, K.; Zhu, K.; Liu, W.J.; Pi, Q.; Byambaa, B.; Dokmeci, M.R.; Shin, S.R.; et al. Direct 3D bioprinting of perfusable vascular constructs using a blend bioink. *Biomaterials* **2016**, *106*, 58–68. [CrossRef] [PubMed]

48. Xu, P.D.; Xie, R.X.; Liu, Y.P.; Luo, G.A.; Ding, M.Y.; Liang, Q.L. Bioinspired microfibers with embedded perfusable helical channels. *Adv. Mater.* **2017**, *29*, 1701664. [CrossRef] [PubMed]

49. Kitagawa, Y.; Naganuma, Y.; Yajima, Y.; Yamada, M.; Seki, M. Patterned hydrogel microfibers prepared using multilayered microfluidic devices for guiding network formation of neural cells. *Biofabrication* **2014**, *6*, 035011. [CrossRef] [PubMed]

50. Sun, T.; Shi, Q.; Huang, Q.; Wang, H.P.; Xiong, X.L.; Hu, C.Z.; Fukuda, T. Magnetic alginate microfibers as scaffolding elements for the fabrication of microvascular-like structures. *Acta Biomater.* **2018**, *66*, 272–281. [CrossRef] [PubMed]

51. Kang, H.W.; Lee, S.J.; Ko, L.K.; Kengla, C.; Yoo, J.J.; Atala, A. A 3D bioprinting system to produce human-scale tissue constructs with structural integrity. *Nat. Biotechnol.* **2016**, *34*, 312–319. [CrossRef] [PubMed]

52. Ghorbanian, S.; Qasaimeh, M.A.; Akbari, M.; Tamayol, A.; Juncker, D. Microfluidic direct writer with integrated declogging mechanism for fabricating cell-laden hydrogel constructs. *Biomed. Microdevices* **2014**, *16*, 387–395. [CrossRef] [PubMed]

53. Li, X.F.; Shi, Q.; Wang, H.P.; Sun, T.; Huang, Q.; Fukuda, T. Magnetically-guided assembly of microfluidic fibers for ordered construction of diverse netlike modules. *J. Micromech. Microeng.* **2017**, *27*, 125014. [CrossRef]

*Article*

# Physical Properties of the Extracellular Matrix of Decellularized Porcine Liver

Hiroyuki Ijima [1,*], Shintaro Nakamura [1], Ronald Bual [1], Nana Shirakigawa [1] and Shuichi Tanoue [2]

[1]   Department of Chemical Engineering, Faculty of Engineering, Graduate School, Kyushu University, Fukuoka 819-0395, Japan; s.nakamura@kyudai.jp (S.N.); ronald.bual@g.msuiit.edu.ph (R.B.); shirakigawa@chem-eng.kyushu-u.ac.jp (N.S.)
[2]   Frontier Fiber Science and Technology, Faculty of Engineering, University of Fukui, Fukui 910-8507, Japan; tanoue@matse.u-fukui.ac.jp
*   Correspondence: ijima@chem-eng.kyushu-u.ac.jp; Tel./Fax: +81-92-802-2748

Received: 7 March 2018; Accepted: 26 April 2018; Published: 1 May 2018

**Abstract:** The decellularization of organs has attracted attention as a new functional methodology for regenerative medicine based on tissue engineering. In previous work we developed an L-ECM (Extracellular Matrix) as a substrate-solubilized decellularized liver and demonstrated its effectiveness as a substrate for culturing and transplantation. Importantly, the physical properties of the substrate constitute important factors that control cell behavior. In this study, we aimed to quantify the physical properties of L-ECM and L-ECM gels. L-ECM was prepared as a liver-specific matrix substrate from solubilized decellularized porcine liver. In comparison to type I collagen, L-ECM yielded a lower elasticity and exhibited an abrupt decrease in its elastic modulus at 37 °C. Its elastic modulus increased at increased temperatures, and the storage elastic modulus value never fell below the loss modulus value. An increase in the gel concentration of L-ECM resulted in a decrease in the biodegradation rate and in an increase in mechanical strength. The reported properties of L-ECM gel (10 mg/mL) were equivalent to those of collagen gel (3 mg/mL), which is commonly used in regenerative medicine and gel cultures. Based on reported findings, the physical properties of the novel functional substrate for culturing and regenerative medicine L-ECM were quantified.

**Keywords:** liver-specific extracellular matrix; gel; scaffold; tissue engineering; decellularization; solubilized extracellular matrix; physical property; storage modulus; loss modulus; biodegradation

---

## 1. Introduction

Tissue engineering is a promising approach in realizing regenerative medicine, manifested by the combined use of cells, scaffolds, and growth factors. In particular, the reconstruction of the liver, which is the center of metabolism in our body, is one of the most important and difficult tasks in regenerative medicine. Herein, the scaffold is considered an exogenous pericellular factor that affects cells. Various materials have been developed as scaffolding materials for tissue engineering, and these have been broadly classified into natural and synthetic polymers. Natural biological macromolecules, such as collagen and laminin, are characterized by excellent cell signal transmission. For example, they are effective materials for the adhesion and proliferation of cells as well as for the expression of differentiation and functions.

In recent years, attention has been paid to decellularized tissues from which cellular components have been completely removed [1–4]. Since they are biologically derived materials, they possess excellent biocompatibility and capacity in maintaining the vascular network of tissues after decellularization.   Therefore, they have been extensively studied as an organ template for transplantation. The recellularization of the decellularized tissue should be optimized in order to fabricate a functional reconstructed liver. Previous reports suggested that recellularized liver inoculated with hepatocytes expresses liver-specific functions such as albumin and urea synthesis [5,6]. Also, site-specific cell inoculation has been successfully performed [7]. However, the hepatocytes density of recellularized liver evaluated from the inoculum cell number was $10^6$–$10^7$ cell/mL. These values are significantly below the level of 1–2 $\times$ $10^8$ cells/mL, which is considered to be the cell density in a healthy liver. In this regard, the formation of new liver tissue accompanied by cell growth within the organ template is essential. Also, recellularized liver transplantation experiments have not been successful for even a few days of cell engraftment [3,8]. Furthermore, in in vivo, cells are embedded in the ECM. Therefore, when the decellularized tissue is used as a scaffold material, it is difficult to reproduce a similar in vivo environment to that which existed before decellularization, where individual cells will be embedded in a similar manner within the ECM. In order to overcome the above problems, liver-specific ECM gel construction is indispensable as a suitable environment in the formation of functional liver tissue accompanied by angiogenesis [9,10]. Meanwhile, functional liver tissue formation accompanied with angiogenesis and hepatocyte proliferation has been reported in vivo by implanting cells with ECM model gel [7,10].

Recently, the maintenance and promotion of the liver phenotype [11–14] was reported to be effective when decellularized liver was solubilized (L-ECM) in the effort to construct a new culture substrate and generate a 3D injectable hydrogel platform for liver tissue engineering [15]. In addition, it is reported that the composition and structure of extracellular matrix (ECM) constituting each organization differs in accordance with the organ type [16,17]. Likewise, it is suggested that organ-specific ECM constitutes an optimal microenvironment around cells. Thus, the solubilized liver-specific ECM (L-ECM) is an effective substrate for different applications, from basic use in drug screening with hepatocyte culture to constructive application in liver regenerative medicine.

The behavior of cells is influenced by the components and structure of the scaffold, and by the material's physical properties. For example, when hepatocytes were embedded using alginate gel with different elastic moduli (1 kPa, 12 kPa), increased albumin synthesis activity was elicited in conditions where the elastic modulus was 1 kPa. In addition, it was reported in the same study that increased albumin synthesis activity was elicited in conditions where the softest PEG-heparin material was used [18]. In other words, material properties are considered to be factors that affect cell behavior. However, cell adhesion, morphology, tissue formation ability, and functional expression are determined by the overall influence of the composition, surface characteristics, and mechanical properties of the substrate. Though there is a report on the mechanical properties of rat L-ECM [15], it is essential to consider the L-ECM of other animals (e.g., porcine) clinically and to examine their physical properties in detail.

For these reasons, it is important to quantify the physical properties of L-ECM. However, detailed investigations on them have not been reported. Therefore, in this study, L-ECM obtained by the solubilization of decellularized porcine liver, as well as L-ECM gel obtained by the spontaneous association of L-ECM, were prepared and their physical properties were evaluated.

## 2. Results and Discussion

### 2.1. Decellularization of Porcine Liver Tissue and Preparation of L-ECM

The immunohistological evaluation of porcine liver slices decellularized by Triton X-100 treatment was performed. In the histological evaluation of the liver after decellularization, the absence of the cellular cytoplasms and nuclei of cellular components was confirmed (Figure 1). Furthermore, since 92.4% of the DNA in the porcine tissue was removed in this treatment, it was confirmed that the decellularization of porcine liver slices by 1% Triton X-100 treatment was successful. Using immunohistological evaluations of this decellularized liver slice, collagen types I, III, IV, and V, and laminin originally contained in the native liver were detected (Figure 2). Despite the qualitative nature of fluorescence imaging, distinct red fluorescence was observed by staining using type I collagen antibodies. In addition, white powder was obtained by the lyophilization of this decellularized liver tissue. These series of processes lasted approximately 15 days and led to the harvesting of 18.0 ± 0.2 mg of decellularized porcine liver dry powder per 1 g (wet weight) of porcine liver tissue. Furthermore, by treating this powder with pepsin, L-ECM was obtained as liver-specific solubilized ECM (Figure 3). This L-ECM contained approximately 0.15 µg/mg of glycosaminoglycan (GAG), which was equivalent to the concentration yield of our previous study [11]. These results indicate that the acquisition of L-ECM as a solubilized substrate had a liver-specific matrix composition since all of the cellular components were removed.

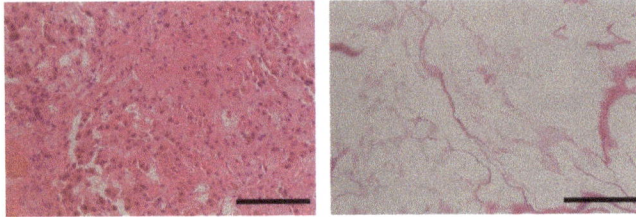

**Figure 1.** Histological observation (Hematoxylin and eosin staining) of native liver (**A**) and decellularized liver (**B**) (scale bars = 100 µm).

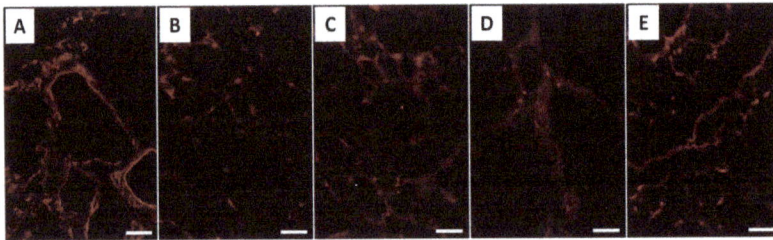

**Figure 2.** Immunolabeling of collagen types (**A**) I, (**B**) III, (**C**) IV, (**D**) V, and laminin (**E**), in decellularized samples with Triton solution. All images are displayed at a magnification of 20 times (scale bars = 50 µm).

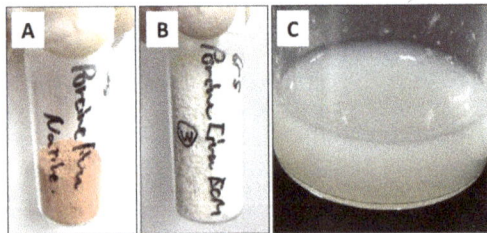

**Figure 3.** (**A**) Appearance of natural liver powder, (**B**) decellularized liver powder, and (**C**) L-ECM. Each powder type was obtained by the lyophilization of the corresponding liver tissue.

*2.2. Characterization of L-ECM*

SDS-PAGE results and the signal ratios for each molecular weight band of all studied solutions are shown in Figure 4. Molecular weight bands were observed at MW = 116 and 227 kDa at all studied conditions (Figure 4A). Additionally, in each solution, differences in the relative densities of each molecular weight band were confirmed. Therefore, as shown in Figure 4B, the proportions of each molecular weight band for all studied molecules were quantified by image analyses using Image J.

In the acid-solubilized porcine type I collagen (I Col) (Nitta Gelatin, Osaka, Japan), the γ and the β chains were the most prominent, while the proportion of the α chains was relatively small. Conversely, in pepsin-solubilized porcine type I collagen (PI Col) (Nitta Gelatin), the proportion of the α chains was large and the ratio of the γ and β chains was small. Additionally, the same trend was observed for L-ECM as that for the PI Col. In other words, it can be said that the L-ECM obtained in this study is a pepsin-solubilized liver-specific matrix substrate.

In the solubilization of collagen by pepsin, telopeptide is cleaved by hydrolysis [19]. Therefore, the γ chains, which is a trimer of the collagen chain and the dimerized β chains, are present in increased quantities in the acid-solubilized collagen. In contrast, the α chain, which is a single strand, exists in pepsin-solubilized collagen in increased quantities. Since L-ECM was pepsin-solubilized, the proportion of the α chains was the most prominent. Therefore, it was expected that pepsin removed telopeptides from the L-ECM. In type I collagen, the triple helix-forming site is composed of glycine–X–Y (X and Y represent other amino acids; proline and hydroxyproline are used in many cases). Conversely, since the telopeptide does not contain this type of repeating sequence and is a species-specific sequence for the animal that is being studied, it becomes the main antigenic site of collagen [19]. Therefore, the removal of telopeptides leads to the suppression of immune reactions so that pepsin-treated L-ECM is expected to become a biocompatible material for realizing tissue engineering technologies.

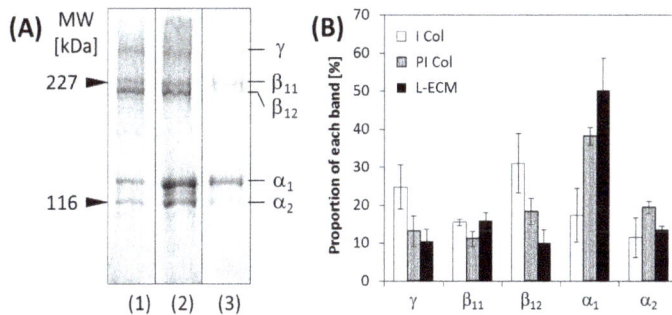

**Figure 4.** I (**A**) SDS-PAGE of each sample. (1) Solubilization with acid (I Col), (2) pepsin-acid solubilization (PI Col), and (3) L-ECM. The concentration of each solution was 0.75 mg/mL. (**B**) Ratios of detected bands in the all components (*n* = 3, bars represent standard deviation).

*2.3. Rheological Properties of L-ECM Based on Dynamic Viscoelastic Evaluation*

All of the rheological measurements were performed in wet conditions.

2.3.1. Strain Dispersion Test

In the strain dispersion test (Figure 5A,B), PI Col yields a constant value within the strain range of 0.1–20%. When the strain amount exceeds 10–20%, $G'$ suddenly drops, and the relative value of $G''$ is increased. It can be judged that the viscous behavior became more prominent. At the concentration of 2 mg/mL or higher, the strain amount at $G' = G''$ (intersection of $G'$ and $G''$) shifted to a higher distortion side as the concentration increased. This indicates that the elastic behavior becomes more prominent as the concentration increases. At 1 mg/mL, no intersection point between $G'$ and $G''$ was observed in the measured strain range. Therefore, the material was considered to be in a relatively viscous state.

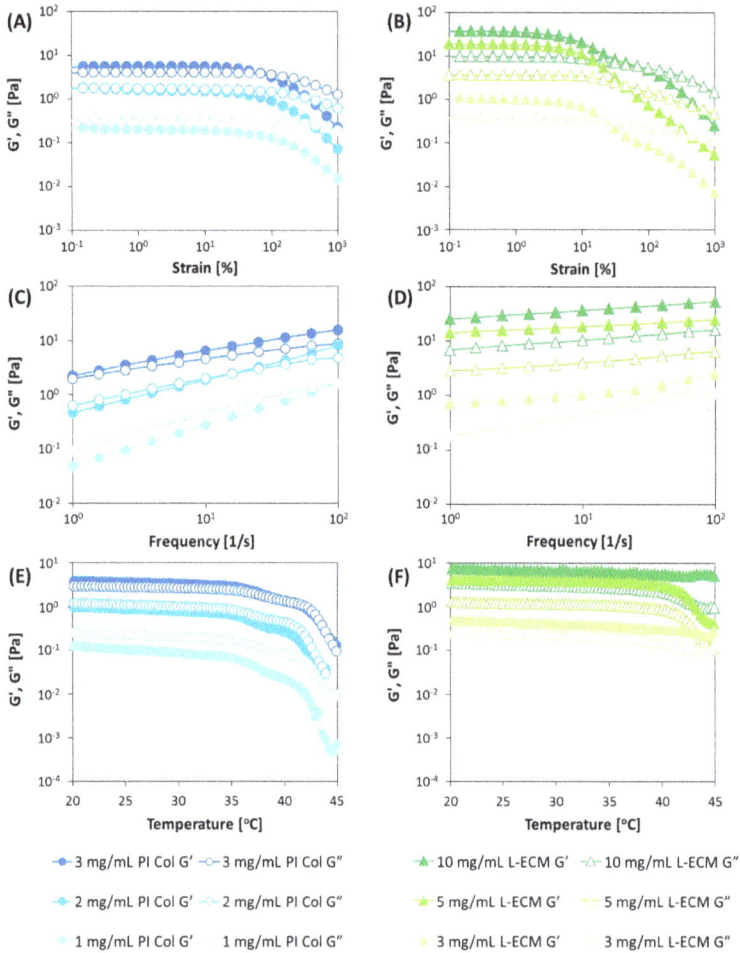

**Figure 5.** (**A,C,E**) Variations in $G'$ and $G''$ at different concentrations of PI Col (3, 2, and 1 mg/mL), (**B,D,F**) L-ECM concentrations (10, 5, and 3 mg/mL). Corresponding variations of $G'$ and $G''$ with (**A,B**) dynamic strain sweep (DSS) mode, (**C,D**) dynamic frequency sweep (DFS) mode, and (**E,F**) dynamic temperature ramp (DTR) mode measured by a rheometer (conditions for DSS: frequency 1 Hz, temperature 10 °C; conditions for DFS: strain 5%, temperature 10 °C; and conditions for DTR: frequency 1 Hz, and strain 5%). All samples were dissolved in an aqueous solution adjusted with HCl at pH = 3.0.

However, L-ECM yielded a constant value in the strain range of 0.1 to 10% and exhibited a linear behavior in a narrower range compared to PI Col. In regard to the strain dispersion, the reduction of $G'$ and $G''$ owing to the increase in the strain is considered to be nonlinear in terms of the material's physical properties simply owing to its complicated internal structure. From Figure 5A,B, it can be observed that the strains exhibit nonlinearity (distortion values at which $G'$ and $G''$ begin to decrease) decreased when the content was low. In this instance, the internal structure remarkably changed the rheological behavior. In other words, it is considered that the material that exhibits a more prominent nonlinearity has $G' > G''$, and the difference between them is larger. As the content increases, both $G'$ and $G''$ increase, but the difference between them decreases. It is thought that this is due to an increase in the number of molecules, rendering the material difficult to deform owing to the secondary intermolecular bonds (such as hydrogen bonds), and as a result of the increased contribution of plastic deformation. Therefore, the constant elastic modulus region in the strain dispersion test justifies the dispersion stability of the substance in the solvent. It was suggested that the PI Col was excellent in terms of its dispersion stability since its stable strain range response spanned approximately 0.1–20% compared to approximate range of 0.1–10% for L-ECM. This indicates that solutes (such as collagen molecules) are difficult to aggregate and dissociate in the solvent. It is considered that the dispersion stability of L-ECM is lower than that of PI Col owing to the fact that L-ECM is not a completely homogeneous liquid but rather represents a state in which small particles are present. This is manifested by the fact that L-ECM is not clear or transparent, even after solubilization, but has a white and cloudy appearance (Figure 3C).

Similar to the PI Col, concentration increases led to increases in values of $G'$ and $G''$, while the amount of strain indicating the intersection of each term shifted at higher strains. Subsequently, a comparison of L-ECM and PI Col at the same concentration showed that PI Col exhibited a more elastic behavior since the value of $G'$ of PI Col was higher than that of L-ECM. That is, at the same concentration, it can be said that the intermolecular friction of L-ECM is smaller than that of PI Col.

### 2.3.2. Frequency Dispersion Test

In the frequency dispersion test, the vibration amplitude was constant and the vibration frequency was varied and measured. Since both samples showed linear behaviors in the vicinity of strains of approximately 5%, in the strain dispersion test, measurements in the frequency dispersion test were carried out at a strain level of 5% (Figure 5C,D).

In the case of the PI Col, increases in the $G'$ and $G''$ values at increasing frequencies were confirmed. The rate of this increase was larger as the concentration decreased. This is thought to be due to an increase in the force applied to the polymeric chain. The increase in the $G'$ and $G''$ values due to this frequency increase show an effect that is similar to the increase in the intermolecular interaction owing to concentration increases. Therefore, at low frequencies, it is considered that the viscoelastic behavior with relatively low intermolecular interactions, corresponding to low concentrations, affects the value of $G'$ and $G''$.

Herein, the frequencies in the case where $G'' > G'$ for concentrations of 1, 2, and 3 mg/mL of the PI Col were in the frequency regions of approximately 100, 10, and 1 Hz, respectively. This can also be explained by the fact that a dominant viscous behavior was elicited at low concentrations and low frequency conditions. That is, in the high frequency range, only short relaxation times affect $G''$. Correspondingly, $G''$ decreased and $G'$ increased. That is, the behavior resembled that of a liquid in the low frequency region and that of a solid in the high frequency region. It was suggested that the shifting of the point of intersection of $G'$ and $G''$ to the high frequency side was due to the shortening of the typical relaxation time of the material.

Conversely, the increase in $G'$ and $G''$ at higher frequencies was also confirmed in the L-ECM. However, $G'$ and $G''$ did not intersect in the measured frequency range. Comparing the PI Col and the L-ECM at the same concentration, no noticeable differences were found between $G'$ and $G''$ with respect to the frequency changes. Both PI Col and L-ECM showed that $G' > G''$ within the measured frequency ranges, thereby suggesting that this referred to a substrate that had a predominantly elastic behavior. However, the elasticity of PI Col was greater than that of L-ECM since there was a stronger intermolecular interaction in the PI Col compared to L-ECM at the same concentration, as also shown in the strain dispersion evaluation.

In alternative terms, the values of $G'$ and $G''$ of PI Col were relatively close to each other, and the $G'$ value of L-ECM yielded a larger value than $G'$, because L-ECM had a higher elasticity compared to that of PI Col. Given the magnitude relationship between the absolute values of $G'$ and $G''$, it seems that the entanglement of molecules on the L-ECM have a larger influence on the flow behavior than that for the PI Col. From the above frequency dispersion, the values of $G'$ and $G''$ are relatively close to each other in the case of the PI Col, and the behavior of the L-ECM is relatively rubbery.

That is, in L-ECM, molecular bonds are formed and the structure is expected to change or be extremely intertwined. In addition, it was suggested that the typical relaxation time of L-ECM is short since the decrease of $G''$ at increasing frequencies was not confirmed in the L-ECM.

### 2.3.3. Temperature Dependence

The temperature dependence (Figure 5E,F) was investigated at a strain of 5% and a frequency of 1 Hz. As a result of the measurement of the $G'$ and $G''$ values in the course of the transition from low to high temperatures, PI Col elicited constant $G'$ and $G''$ values as a function of temperature, but began to decrease after a certain temperature. This decline began at the porcine collagen denaturation temperature of 37 °C. It seems that the physical properties of the material changed due to protein denaturation. However, $G'$ and $G''$ decreased gradually with temperature increases in the case of L-ECM, and a sudden decrease in the elastic modulus was observed at the same temperature as that observed in the case of the PI Col.

Based on the dynamic viscoelasticity evaluation of the L-ECM solution, it was confirmed that its elastic behavior became prominent and dominant in a concentration-dependent manner. In addition, this elicited behavior can be primarily attributed to the small number of molecular associations, and secondarily attributed to the weak intermolecular forces of the L-ECM constituent components. In the ImageJ analyses of the SDS-PAGE results, the proportion of α chains was larger in PI Col than in I Col, and the proportion of α chains was larger in L-ECM than in PI Col. The removal of the telopeptide decreased the fibrillogenic ability of collagen [20]. Based on this finding, it is considered that in the L-ECM there are few molecules associated with collagen chains, and the entanglement between molecules is poor. Secondly, it is a multicomponent mixed system. It has been reported that molecules, such as proteoglycans and fibronectin, affect the fibril formation of collagen due to their coexistence with collagen [21–23]. Thus, it is suggested that repulsion between molecules exists owing to the multicomponent system.

### 2.4. SEM Gel Observations

SEM images of L-ECM and the I Col gel are shown in Figure 6. Since these photos were obtained by dehydration treatment, there is no guarantee that they accurately indicate the state of the ECM in the actual hydrogel. However, these images are important data to infer the skeleton structure of ECM gel [15,24]. Fibrous skeleton was observed in all samples. When comparing (A) the 3 mg/mL I Col gel and (B) the 10 mg/mL L-ECM gel samples, the densities of the fibers constituting the gels are comparable, whereas the fiber diameter in the 3 mg/mL I Col gel sample was thicker than that in the 10 mg/mL L-ECM gel sample. In addition, when comparing (B) the 10 mg/mL L-ECM and (C) the 20 mg/mL L-ECM gel samples, the fiber diameter was comparable. However, the fiber density was higher in the 20 mg/mL L-ECM gel sample compared to the 10 mg/mL L-ECM gel sample. Therefore, the diameters of the fibers constituting the gel of the 10 mg/mL L-ECM gel sample were smaller than those of the 3 mg/mL I Col gel sample, but the fiber densities were almost equal. In addition, the fiber density of the 20 mg/mL L-ECM gel was the highest at the conditions at which the tests were conducted. In other words, it was quantitatively found that the skeleton fiber diameter depends on the constituent components, and the fiber density depends on the concentration. In addition, these differences may be the result of other ECM components affecting the spontaneous self-assembly of type I collagen in L-ECM.

Spheroid, a spherical aggregate formed by the assembly of a large number of hepatocytes, expresses relatively better liver-specific functions [25,26]. However, hepatocytes embedded in spontaneously self-assembled collagen gel express liver functions that are equivalent to those of spheroids, even in the dispersed single cell state [9]. Furthermore, collagen gel-embedded hepatocyte spheroids express liver functions that are much higher than the above [9]. In other words, the in vivo-like self-assembled collagen fiber network is meaningful for cell culture and related tissue engineering applications. Therefore, L-ECM [11] gel, which has a synergistic effect with this collagen fiber network and other ECM components (such as GAGs capable of growth factor immobilization), is expected to be an important substrate promoting functional liver tissue formation.

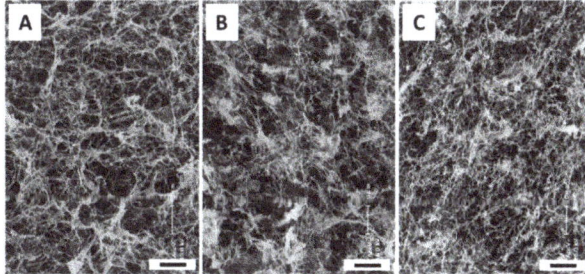

**Figure 6.** (**A**) SEM images of 3 mg/mL I Col gel, (**B**) 10 mg/mL L-ECM gel, and (**C**) 20 mg/mL L-ECM gel samples (all images are shown at a magnification of 1000 times. Scale bar = 10 μm).

*2.5. Immunostaining of L-ECM Gel*

Immunostaining images of L-ECM gel are shown in Figure 7. Fluorescence was confirmed only in the case where the type I collagen antibody was used in the I Col gel sample. Even if differences in fluorescence intensity were observed, red fluorescence was observed in the L-ECM gel when any antibody was used. Among all of the stained images, the strongest intensity in red fluorescence was detected when the type I collagen antibody was used. In other words, it was suggested that the L-ECM gel is a functional gel substratum that consisted of the ECM components that were originally present in the liver.

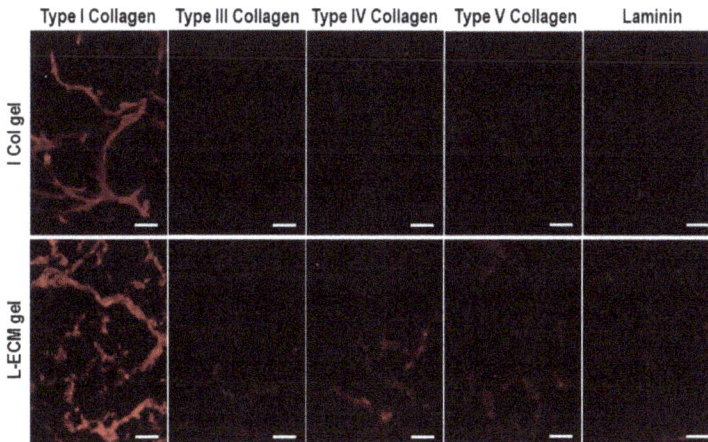

**Figure 7.** Fluorescent immunostaining images of I Col and L-ECM gel (scale bar = 50 μm).

*2.6. Rheological Properties of Gelation Behavior of L-ECM*

The gelation behaviors of L-ECM and I Col are shown in Figure 8. For all of the studied conditions, an increase in the elastic modulus was confirmed as the temperature increased. In addition, the loss elastic modulus never exceeded the storage modulus within the measured temperature range. Here, similar gelation behaviors were observed between 10 mg/mL L-ECM and 3 mg/mL I Col solutions. In fact, in the preliminary study using rat L-ECM, the content of type I collagen was only about half of L-ECM (data not shown). Though there may be some effects of differences in species, their similarity in gelation kinetics can be deduced to be reasonable.

Normally, in the solution state, the loss elastic modulus yields a value higher than the storage modulus. Conversely, in the gel, the storage modulus exceeds the loss elastic modulus. This is due to the change to a solid state by gelation, whereby externally applied energy is stored and converted to a repulsive force. However, the earliest gel formation for L-ECM started at a concentration of 20 mg/mL. This is thought to be due to the frequent entanglement of components owing to an increase in the concentration and relatively early initiation of nucleation of fibers that occur at the initial stage of fibril formation.

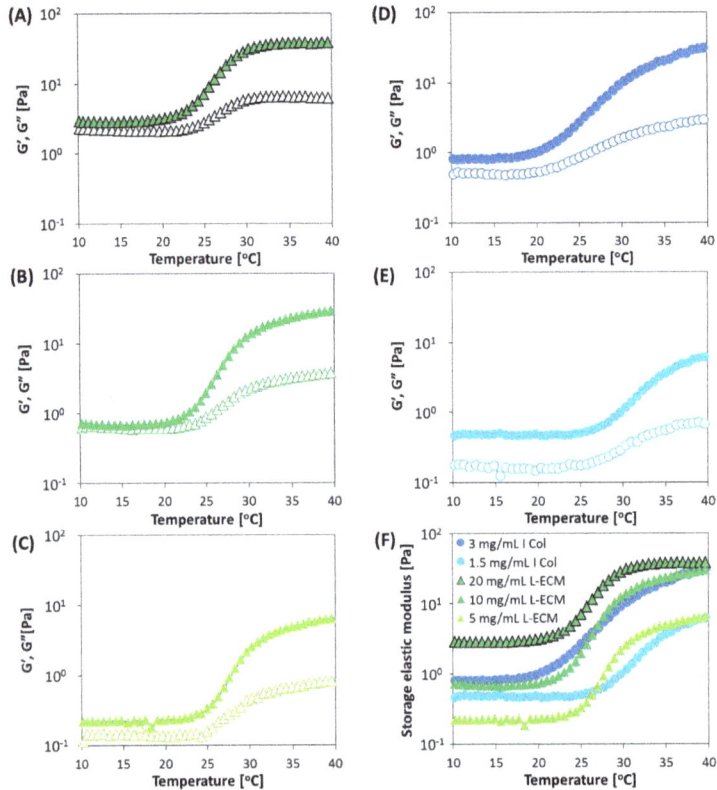

**Figure 8.** Change in $G'$ and $G''$ at different concentrations of I Col (3 and 1.5 mg/mL) and L-ECM solutions (20, 10, and 5 mg/mL) with the rheometer measurements conducted in a DTR mode. (**A**) 20 mg/mL L-ECM, (**B**) 10 mg/mL L-ECM, (**C**) 5 mg/mL L-ECM, (**D**) 3 mg/mL I Col, (**E**) 1.5 mg/mL I Col, and (**F**) storage modulus for all tested conditions. (**A–E**) Open and closed symbols indicate storage and loss moduli, respectively (conditions of DTR: frequency, 1 Hz, and strain, 5%. Temperature was increased at 2 °C /min from 10 to 40 °C. The pH of all samples was adjusted to its value in physiological conditions).

## 2.7. Degradation Behavior of L-ECM Gel

The decomposition behaviors of L-ECM and I Col gel by collagenase are shown in Figure 9. Samples underwent digestion with collagenase at all tested conditions, and a decrease in gel weight over time was confirmed. In addition, solutions of 10 mg/mL L-ECM and 3 mg/mL I Col gel showed comparable degradation rates. Since L-ECM is composed of atelocollagen from which telopeptide had been removed, it can be easily degraded by proteases. The condition with the highest degradation rate was that for I Col gel with a concentration of 1.5 mg/mL, and the condition with the slowest degradation rate was that for L-ECM gel with a concentration of 20 mg/mL.

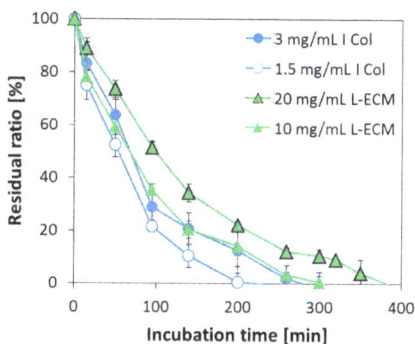

**Figure 9.** Degradation profiles of I Col gels (3 and 1.5 mg/mL) and L-ECM gels (20 and 10 mg/mL) using collagenase solution at a concentration of 0.05 mg/mL as a function of incubation time ($n = 3$, bars represent standard deviation).

### 2.8. Stress on Compression of L-ECM Gel

The stress-strain curves of L-ECM and I Col gel are shown in Figure 10, and the upper yield point and elastic modulus of each sample are shown in Table 1. The compression stress was detected at all tested conditions. In addition, a concentration-dependent compression response was obtained with I Col and L-ECM. Comparison of the upper yield points of I Col and L-ECM at each concentration yielded a maximum at 3 mg/mL for I Col, and its value was 0.0112 N/mm$^2$. The elastic modulus was also the highest at the same conditions and its value was 0.0825 N/mm$^2$. Comparing the I Col and L-ECM gels at the same concentration (3 mg/mL), the upper yield point and elastic modulus of I Col were 0.0112 and 0.0825 N/mm$^2$, respectively, while those of L-ECM were 0.0008 and 0.0057 N/mm$^2$, respectively. Furthermore, the gel compression test showed that the L-ECM gel with a concentration of 10 mg/mL elicited an inferior mechanical strength response compared to 3 mg/mL I Col.

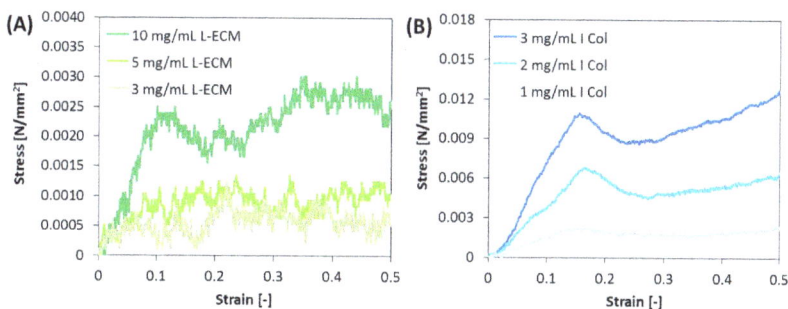

**Figure 10.** Average stress-strain responses ($n = 3$) of (**A**) L-ECM gel, and (**B**) I Col gel.

Table 1. Upper yield point and elastic modulus of I Col and L-ECM gels.

| Sample | Concentration (mg/mL) | Upper Yield Point (N/mm$^2$) | Elastic Modulus (N/mm$^2$) |
|--------|--------|--------|--------|
| I Col | 3 | 0.0112 | 0.0825 |
|  | 2 | 0.0069 | 0.0428 |
|  | 1 | 0.0025 | 0.0175 |
| L-ECM | 10 | 0.0026 | 0.0318 |
|  | 5 | 0.0014 | 0.0153 |
|  | 3 | 0.0008 | 0.0057 |

The fiber diameters constituting the L-ECM gel were smaller than those for the I Col gel, as confirmed by SEM imaging. The degradation and gelling behaviors of the L-ECM gel were equivalent to those of I Col with a concentration of 3 mg/mL, but the mechanical strength of the gel was low. The reason for this is considered to be the fineness of the fiber diameters constituting the gel. In the rheological evaluation, the gelation behavior of the L-ECM gel sample with a concentration of 10 mg/mL was equivalent to that of the I Col gel sample with a concentration of 3 mg/mL. In general terms, the information obtained regarding the gelling behavior does not indicate the strength of the gel [27]. Therefore, through the comparison of the rheological evaluation results that showed the gelation behavior, a well as the compression test results that showed the mechanical strength, the L-ECM was found to elicit a gelation behavior similar to that of I Col, but the strength of the L-ECM gel was lower than that of the I Col gel.

However, it is possible to increase the mechanical strength. For example, this can be achieved using the EDC/NHS reaction in which ethyl (dimethylaminopropyl) carbodiimide (EDC) and N-hydroxysuccinimide (NHS) are used in combination [28]. In addition, collagen gel with high biocompatibility and excellent mechanical strength can be obtained by crosslinking the gel by transglutaminase [29,30]. It is also possible to manipulate the fiber density of the gel skeleton by increasing the concentration of the solution.

## 3. Conclusions

L-ECM was prepared as a liver-specific matrix substrate from the decellularized porcine liver. It had the characteristics of a pepsin-digested substrate and retained the various components contained in the native liver. L-ECM had a lower elasticity compared to pepsin-digested type I collagen, and showed an abrupt decrease in its elastic modulus at 37 °C. The elastic modulus increased with increasing temperatures (up to 40 °C), and the loss elastic modulus never exceeded the storage elastic modulus in the L-ECM gel. In addition, as the L-ECM gel concentration increased, decreases in the biodegradation rate and increases in the mechanical strength were confirmed. All of the elicited properties for L-ECM at a concentration of in 10 mg/mL were equivalent to collagen gel at a concentration of 3 mg/mL, and this concentration of collagen is commonly used for regenerative medicine and gel cultures.

It is expected that the obtained results will greatly contribute to the optimization of the scaffold for hepatic tissue engineering. Furthermore, the developed L-ECM is expected to be used as a substrate for functional hepatocyte culture. Lastly, the results will provide important information for understanding the liver-specific phenotypic expression of hepatocytes and formed liver tissue.

## 4. Materials and Methods

### 4.1. Decellularization of Porcine Liver

A healthy porcine liver (Kyudo, Saga, Japan) was depleted of blood in a fresh state with calcium and magnesium-free phosphate-buffered saline (CMF–PBS) and was cryopreserved at −80 °C until use. The porcine liver (Kyudo, Tosu, Japan) was sectioned in slices with a thickness of 2 mm and was decellularized by stirring at 4 °C with 1% Triton-X 100 (polyoxyethylene-p-isooctylphenol) (Sigma-Aldrich, St. Louis, MO, USA) in CMF-PBS. Thereafter, the decellularized liver was washed with CMF-PBS and further dialyzed against water. The dialysis membrane used was Spectra/Por 6 (MWCO: 1000, Spectrum Laboratories, Inc., Rancho Dominguez, CA, USA). After washing, lyophilization treatment was applied to obtain dried porcine liver ECM. The animal experimental protocol was reviewed and approved by the Ethics Committee on Animal Experiments of Kyushu University (A25-282-0, 21 Feb 2014).

### 4.2. Preparation of Liver-Specific ECM-Solubilized Substrate and Preparation of Hydrogel

Approximately 10 mg of lyophilized decellularized liver was placed in 1 mL of pepsin (Sigma-Aldrich) solution (1 mg/mL in 0.1 N HCl). Solubilized ECM derived from decellularized liver was obtained by the treatment with pepsin solution at 4 °C for 72 h. The pH of the ECM was adjusted to pH 3.0 by dialysis with Spectra/Por 6 (MWCO: 1000, Spectrum Laboratories, Inc.), and solubilized liver-specific ECM was obtained (L-ECM). DNA and glycosaminoglycan (GAG) contents were quantitatively analyzed using a Fluorescent DNA Quantification Kit (Bio-Rad Laboratories Inc., Hercules, CA, USA) and GAGs quantitative kit (Euro Diagnostica AB, Malmö, Sweden) assay according to the recommended protocol. L-ECM was mixed with concentrated minimum essential medium eagle (MEM) ($\times$10) and buffer (47.7 mg HEPES/mL, 0.08N NaOH) at a volume ratio of 8:1:1 ($v/v$) and was kept on ice. The solution formed a gel after incubation at 37 °C for 30 min by assembling itself into a three-dimensional network.

### 4.3. Immunostaining

Sections with a thickness of 10 μm were prepared using a freezing microtome (CM 1100: Leica Microsystems GmbH, Wetzlar, Germany) and immobilized by immersion in 4% formaldehyde for 10 min. It was then washed with CMF-PBS and was immersed in CMF-PBS that was supplemented with 1% bovine serum albumin (BSA) (Wako, Osaka, Japan) for 30 min for blocking treatment. Primary antibody diluted with CMF-PBS was supplemented with 1% BSA was then added in a dropwise manner to the sample. It was left overnight at 4°C. After washing with CMF-PBS supplemented with 1% BSA, a secondary antibody was diluted with CMF-PBS that was supplemented with 1% BSA. It was added drop-by-drop to the sample and was allowed to stand at room temperature for 1 h. All operations after the addition of the secondary antibody were conducted using light shielding. Subsequently, the sample was washed with CMF-PBS supplemented with 1% BSA, covered with a cover glass, and observed with a fluorescence microscope. Details of the antibodies used were as follows. All primary antibodies were rabbit-derived antibodies against rat proteins. Anti-collagen type I was purchased from Rockland Antibodies and Assays (Limerick, PA, USA). Anti-collagen type III, anti-collagen type IV, and anti-collagen type V were purchased from Abbiotech (San Diego, CA, USA). Anti-laminin was purchased from Bioss Antibodies (Woburn, MA, USA). In addition, goat anti-rabbit IgG TRITC-conjugated antibody (ThermoFisher Scientific-Invitrogen, Waltham, MA, USA) was used as a secondary antibody.

### 4.4. Molecular Weight Distribution of L-ECM

SDS-PAGE was used for the investigation of molecular weight distributions. Electrophoresis samples of L-ECM at a concentration of 0.75 mg/mL, acid-solubilized porcine type I collagen (I Col), and pepsin-solubilized porcine type I collagen (PI Col) (Nitta Gelatin Inc.) were prepared. A 5% acrylamide gel was prepared and each sample was electrophoresed at 200 V with 40 mA. Furthermore, by analyzing the SDS-PAGE image obtained by ImageJ, the proportion of each molecular weight in all molecules was calculated.

### 4.5. SEM Gel Observations

Samples of L-ECM at concentrations of 10 and 20 mg/mL as well as I Col samples at concentrations of 3 mg/mL-type were adjusted to neutral pH by mixing with reconstitution buffer (47.7 mg HEPES/mL, 0.08 N NaOH) and ×10 MEM at a ratio of 8:1:1. Each adjusted sample had a concentration that was equal to four-fifths of the above concentration. The solution formed a gel after incubation at 37 °C for 30 min by assembling itself into a three-dimensional network. The sample was substituted for ethanol and t-butanol (Wako) for dehydration. After replacement, the liver was allowed to stand at 4 °C, and was then dried using a vacuum pump. The morphological structure of the obtained dried sample was observed with a scanning electron microscope (SEM, SS 550: Shimadzu Co., Kyoto, Japan).

### 4.6. Rheology of L-ECM

Pepsin-solubilized I col (PI Col) samples at concentrations of 1, 2, and 3 mg/mL were used as the relative evaluation conditions for L-ECM samples at concentrations of 3, 5, and 10 mg/mL. Additionally, the apparatus used was a viscoelasticity measuring apparatus MCR (Anton Paar, Graz, Austria). In the analysis, the storage modulus ($G'$, elastic term) and loss modulus ($G''$, viscous term) were determined and expressed as respective graphs with respect to distortion, frequency, and temperature. (1) Distortion dispersion (dynamic strain sweep method, DSS) was measured with a cone plate (0.5°, diameter 50 mm), and the strain was changed from 0.1 to 1000% with a frequency of 1 Hz and a temperature of 10 °C for these measurements. (2) Frequency dispersion (dynamic frequency sweep method, DFS) was measured using a cone plate (0.5°, diameter 50 mm) by changing the frequency from 1 to 100 Hz with a strain of 5% and a temperature of 10 °C. (3) Temperature dispersion (dynamic temperature ramping method, DTR) was obtained by increasing the temperature from 10 to 50 °C at 1 °C/min using a parallel plate (50 mm in diameter) at a frequency of 1 Hz and a strain of 5%.

In the preliminary study, L-ECM prepared using powders stored at room temperature for 1 week gelled promptly under neutral conditions. However, when powders stored at 37 °C were used, the gelling ability of L-ECM remarkably decreased (data not shown). In other words, L-ECM before gelation was affected by insufficient stability at body temperature. Therefore, in (1) and (2), measurement was carried out at 10 °C.

### 4.7. Rheology of Gelation Behavior of L-ECM

In order to investigate the gelation behavior of L-ECM, rheological evaluation was performed. Samples of L-ECM at the concentrations of 5, 10, and 20 mg/mL, as well as I Col samples at the concentrations of 1.5 and 3 mg/mL, were adjusted to neutral pH by mixing with reconstitution buffer and ×10 MEM at a ratio of 8:1:1. The storage elastic moduli of these samples were measured by linearly increasing the temperature from 10 to 40 °C at 2 °C/min using a rheometer at a frequency of 1 Hz. A parallel plate (diameter: 50 mm) was used for the measurement.

### 4.8. Degradation Behavior of L-ECM Gel

Collagenase digestion was performed to investigate the degradation properties of the gel. Samples of L-ECM at the concentrations of 10, 20 mg/mL, or I Col samples at the concentrations of 1.5, 3 mg/mL, were adjusted to neutral pH by mixing them with reconstitution buffer and ×10 MEM at a ratio of 8:1:1. Each adjusted sample had a concentration that was equal to four-fifths of the above concentration. A gel volume of 500 μL was shaken in 10 mL of 0.5 mg/mL collagenase (Wako)/0.05 mg/mL trypsin inhibitor (Wako) solution mixture, and the residual weight of the gel was measured as a function of time. The difference from the initial weight was calculated and evaluated as the manifestation of its degradation characteristics.

*4.9. Stress on the Compression of L-ECM Gel*

The mechanical properties of the gel against compression were evaluated. The mechanical strength of the gel was evaluated by determining the stress-strain curve, and by investigating the stress at the upper yield point obtained by increasing the load beyond the elastic limit, thereby identifying the plastic deformation point of the gel. In addition, the elastic modulus was obtained from the linearly increasing parts of the stress-strain curves. Samples of L-ECM solutions with concentrations of 3, 5, and 10 mg/mL, or I Col samples with concentrations of 1, 2, and 3 mg/mL, were added to a 96-well plate at volumes of 300 μL/well and were incubated at 37 °C overnight. The stress on the compression of the gel was measured using a load measuring machine (LTS-50N-S100: Minebea Co., Nagano, Japan). The height of the gel was measured and a stress-strain curve was generated.

**Author Contributions:** S.N. and H.I. conceived and designed the experiments; S.N. performed the experiments; S.N. and S.T. analyzed the rheological properties; H.I., R.B., and N.S. wrote the paper.

**Conflicts of Interest:** The authors declare no conflict of interest.

## References

1. Sasaki, S.; Funamoto, S.; Hashimoto, Y.; Kimura, T.; Honda, T.; Hattori, S.; Kobayashi, H.; Kishida, A.; Mochizuki, M. In vivo evaluation of a novel scaffold for artificial corneas prepared by using ultrahigh hydrostatic pressure to decellularize porcine corneas. *Mol. Vis.* **2009**, *15*, 2022–2028. [PubMed]
2. Poh, M.; Boyer, M.; Solan, A.; Dahl, S.L.; Pedrotty, D.; Banik, S.S.; McKee, J.A.; Klinger, R.Y.; Counter, C.M.; Niklason, L.E. Blood vessels engineered from human cells. *Lancet* **2005**, *365*, 2122–2124. [CrossRef]
3. Uygun, B.E.; Soto-Gutierrez, A.; Yagi, H.; Izamis, M.L.; Guzzardi, M.A.; Shulman, C.; Milwid, J.; Kobayashi, N.; Tilles, A.; Berthiaume, F.; et al. Organ reengineering through development of a transplantable recellularized liver graft using decellularized liver matrix. *Nat. Med.* **2010**, *16*, 814–820. [CrossRef] [PubMed]
4. Baptista, P.M.; Siddiqui, M.M.; Lozier, G.; Rodriguez, S.R.; Atala, A.; Soker, S. The use of whole organ decellularization for the generation of a vascularized liver organoid. *Hepatology* **2011**, *53*, 604–617. [CrossRef] [PubMed]
5. Butter, A.; Aliyev, K.; Hillebrandt, K.H.; Raschzok, N.; Kluge, M.; Seiffert, N.; Tang, P.; Napierala, H.; Muhamma, A.I.; Reutzel-Selke, A.; et al. Evolution of graft morphology and function after recellularization of decellularized rat livers. *J. Tissue Eng. Regen. Med.* **2018**, *12*, e807–e816. [CrossRef] [PubMed]
6. Kojima, H.; Yasuchika, K.; Fukumitsu, K.; Ishii, T.; Ogiso, S.; Miyauchi, Y.; Yamaoka, R.; Kawai, T.; Katayama, H.; Yoshitoshi-Uebayashi, E.Y.; et al. Establishment of practical recellularized liver graft for blood perfusion using primary rat hepatocytes and liver sinusoidal endothelial cells. *Am. J. Transplant.* **2018**. [CrossRef] [PubMed]
7. Shirakigawa, N.; Takei, T.; Ijima, H. Base structure consisting of an endothelialized vascular-tree network and hepatocytes for whole liver engineering. *J. Biosci. Bioeng.* **2013**, *116*, 740–745. [CrossRef] [PubMed]
8. Pan, J.; Yan, S.; Gao, J.J.; Wang, Y.Y.; Lu, Z.J.; Cui, C.W.; Zhang, Y.H.; Wang, Y.; Meng, X.Q.; Zhou, L.; et al. In-vivo organ engineering: Perfusion of hepatocytes in a single liver lobe scaffold of living rats. *Int. J. Biochem. Cell Biol.* **2016**, *80*, 124–131. [CrossRef] [PubMed]
9. Ijima, H. Practical and functional culture technologies for primary hepatocytes. *Biochem. Eng. J.* **2010**, *48*, 332–336. [CrossRef]
10. Shirakigawa, N.; Ijima, H. Nucleus number in clusters of transplanted fetal liver cells increases by partial hepatectomy of recipient rats. *J. Biosci. Bioeng.* **2013**, *115*, 568–570. [CrossRef] [PubMed]
11. Nakamura, S.; Ijima, H. Solubilized matrix derived from decellularized liver as a growth factor-immobilizable scaffold for hepatocyte culture. *J. Biosci. Bioeng.* **2013**, *116*, 746–753. [CrossRef] [PubMed]
12. Loneker, A.E.; Faulk, D.M.; Hussey, G.S.; D'Amore, A.; Badylak, S.F. Solubilized liver extracellular matrix maintains primary rat hepatocyte phenotype in-vitro. *J. Biomed. Mater. Res. A* **2016**, *104*, 957–965. [CrossRef] [PubMed]
13. Saheli, M.; Sepantafar, M.; Pournasr, B.; Farzaneh, Z.; Vosough, M.; Piryaei, A.; Baharvand, H. Three-Dimensional Liver-derived Extracellular Matrix Hydrogel Promotes Liver Organoids Function. *J. Cell Biochem.* **2018**, *119*, 4320–4333. [CrossRef] [PubMed]

14. Damania, A.; Kumar, A.; Teotia, A.K.; Kimura, H.; Kamihira, M.; Ijima, H.; Sarin, S.K.; Kumar, A. Decellularized liver matrix-modified cryogel scaffolds as potential hepatocyte carriers in bioartificial liver support systems and implantable liver constructs. *ACS Appl. Mater. Interfaces* **2018**, *10*, 114–126. [CrossRef] [PubMed]

15. Lee, J.S.; Shin, J.; Park, H.M.; Kim, Y.G.; Kim, B.G.; Oh, J.W.; Cho, S.W. Liver extracellular matrix providing dual functions of two-dimensional substrate coating and three-dimensional injectable hydrogel platform for liver tissue engineering. *Biomacromolecules* **2014**, *15*, 206–218. [CrossRef] [PubMed]

16. Zhang, Y.; He, Y.; Bharadwaj, S.; Hammam, N.; Carnagey, K.; Myers, R.; Atala, A.; Dyke, M.V. Tissue-specific extracellular matrix coatings for the promotion of cell proliferation and maintenance of cell phenotype. *Biomaterials* **2009**, *30*, 4021–4028. [CrossRef] [PubMed]

17. DeQuach, J.A.; Lin, J.E.; Cam, C.; Hu, D.; Salvatore, M.A.; Sheikh, F.; Christman, K.L. Injectable skeletal muscle matrix hydrogel promotes neovascularization and muscle cell infiltration in a hindlimb ischemia model. *Eur. Cells Mater.* **2012**, *23*, 400–412. [CrossRef]

18. You, J.; Park, S.A.; Shin, D.S.; Patel, D.; Raghunathan, V.K.; Kim, M.; Murphy, C.J.; Tae, G.; Revzin, A. Characterizing the effects of heparin gel stiffness on function of primary hepatocytes. *Tissue Eng. Part A* **2013**, *19*, 2655–2663. [CrossRef] [PubMed]

19. Bairati, A.; Garrone, R. *Biology of Invertebrate and Lower Vertebrate Collagen*; Plenum Press: New York, NY, USA, 1985; ISBN 978-1-4684-7638-5. [CrossRef]

20. Yoshimura, K.; Terashima, M.; Hozan, D.; Shirai, K. Preparation and Dynamic Viscoelasticity Characterization of Alkali-Solubilized Collagen from Shark Skin. *J. Agric. Food Chem.* **2000**, *48*, 685–690. [CrossRef] [PubMed]

21. Payne, A.R.; Whittaker, R.E. Low strain dynamic properties of filled rubbers. *Rubber Chem. Technol.* **1971**, *44*, 440–478. [CrossRef]

22. Freakly, P.K.; Payne, A.R. *Theory and Practice of Engineering with Rubber*; Applied Science Publishers LTD: London, UK, 1978; ISBN 978-0853347729.

23. Snowdend, J.M.; Swann, D.A. The formation and thermal stability of in vitro assembled fibrils from acid-soluble and pepsin-treated collagens. *Biochem. Biophys. Acta* **1979**, *580*, 372–381.

24. Lang, R.; Stern, M.M.; Smith, L.; Liu, Y.; Bharadwaj, S.; Liu, G.; Baptista, P.M.; Bergman, C.R.; Soker, S.; Yoo, J.J.; et al. Three-dimensional culture of hepatocytes on porcine liver tissue-derived extracellular matrix. *Biomaterials* **2011**, *32*, 7042–7052. [CrossRef] [PubMed]

25. Ijima, H.; Matsushita, T.; Nakazawa, K.; Fujii, Y.; Funatsu, K. Hepatocyte spheroids in polyurethane foams: Functional analysis and application for a hybrid artificial liver. *Tissue Eng.* **1998**, *4*, 213–226. [CrossRef]

26. Ijima, H.; Nakazawa, K.; Mizumoto, H.; Matsushita, T.; Funatsu, K. Formation of a spherical multicellular aggregate (spheroid) of animal cells in the pores of polyurethane foam as a cell culture substratum and its application to a hybrid artificial liver. *J. Biomater. Sci. Polym. Ed.* **1998**, *9*, 765–778. [CrossRef] [PubMed]

27. Yoshimura, K.; Chonan, Y.; Shirai, K. Reactivity of Shark, Pig, and Bovine Skin Collagens with Formaldehyde and Basic Chromium Sulfate. *Anim. Sci. Technol. (Jpn.)* **1997**, *68*, 285–292. [CrossRef]

28. Liao, J.; Joyce, E.M.; Sacks, M.S. Effects of decellularization on the mechanical and structural properties of the porcine aortic valve leaflet. *Biomaterials* **2008**, *29*, 1065–1074. [CrossRef] [PubMed]

29. Piez, K.A. *Extracellular Matrix Biochemistry*; Piaz, K.A., Reddi, A.H., Eds.; Elsevier: New York, NY, USA, 1988; pp. 9–11, ISBN 978-0444007995.

30. Bond, M.D.; Van Wart, H.E. Characterization of the individual collagenases from Clostridium histolyticum. *Biochemistry* **1984**, *23*, 3085–3091. [CrossRef] [PubMed]

MDPI

St. Alban-Anlage 66

4052 Basel

Switzerland

Tel. +41 61 683 77 34

Fax +41 61 302 89 18

www.mdpi.com

*Gels* Editorial Office

E-mail: gels@mdpi.com

www.mdpi.com/journal/gels